Ecological Understanding:
The Nature of Theory
and the Theory of Nature

Second Edition

Ecological Understanding:
The Nature of Theory
and the Theory of Nature

Second Edition

Steward T. A. Pickett, Jurek Kolasa, and Clive G. Jones

ELSEVIER

AMSTERDAM • BOSTON • HEIDELBERG • LONDON
NEW YORK • OXFORD • PARIS • SAN DIEGO
SAN FRANCISCO • SINGAPORE • SYDNEY • TOKYO

Butterworth-Heinemann is an imprint of Elsevier

Academic Press is an imprint of Elsevier
30 Corporate Drive, Suite 400, Burlington, MA 01803, USA
525 B Street, Suite 1900, San Diego, California 92101-4495, USA
84 Theobald's Road, London WC1X 8RR, UK

 This book is printed on acid-free paper.

Library of Congress Cataloging-in-Publication Data
Pickett, Steward T., 1950–
    Ecological understanding / S.T.A. Pickett, Jurek Kolasa, and Clive G. Jones. — 2nd ed.
     p.   cm.
    ISBN 978-0-12-554522-8
    1. Ecology–Philosophy.   2. Ecology–Methodology.   I. Kolasa, Jurek.
II. Jones, Clive G.   III. Title.
    QH540.5.P5 2007
    577–dc22

                                                                            2007005796

British Library Cataloguing-in-Publication Data
A catalogue record for this book is available from the British Library.

ISBN: 978-0-12-554522-8

For information on all Academic Press publications
visit our Web site at www.books.elsevier.com

Printed in the United States of America

07   08   09   10   9   8   7   6   5   4   3   2   1

CONTENTS

PREFACE TO THE FIRST EDITION

We wrote this book to share with other ecologists what we have learned about the structure and use of theory and its relationship to the myriad activities that constitute modern science. Our own quest was motivated by the sometimes unclear way in which the term "theory" is used in both scientific publications and informal discussions. We needed to find out what theory was and how it was built. We also wanted to evaluate the varied and often contradictory claims made about what constitutes proper scientific practice. Is prediction really the highest or only goal of science? How might it relate to other activities in which scientists engage?

We began with a series of readings and discussions that fortuitously included works describing the tumult in the modern philosophy of science. This process was tough going for us ordinary scientists, and the concepts took a long time to fathom, but eventually a picture began to emerge that we thought would be valuable for the discipline of ecology. We do not pretend to have become philosophers in the process. In fact, what we have learned and can present here is only a sampling of the wide, deep, and swift stream of the philosophy of science. However, we do attempt to draw our insights together into a coherent picture relevant to ecology. This book is a system of ideas about the philosophy of science by practicing ecologists for practicing ecologists. We beg the forbearance of any philosophers who may encounter it.

We have taken advantage of the current spirit of ecological integration. Ecology deals in novel discoveries, establishing new contexts for existing information, and integrating both into established knowledge. These various endeavors are usually practiced within a suite of disparate specialties, and yet more and more ecologists seem to be willing to cross disciplinary boundaries and levels of organization. The syntheses and unification that might ultimately result from such migration and cross-fertilization have the possibility to revolutionize ecology. The new philosophical understanding of theory and its use may help provide a framework in which integration can be nurtured. Thus, integration is a central theme of this book.

In order to think about how integration can be accomplished, we begin with an overview of understanding, relate that to the structure and dynamics of theory, and indicate how changes in understanding relate to integration in ecology. We also examine the nature of large paradigms that affect ecological integration and the social constraints and contexts of ecological understanding and integration. We end with a discussion of some of the important ways in which ecological understanding intersects with the larger society. In a sense, the book has a symmetrical structure motivated by the need for integration. We begin with a look at the nature of understanding and the tools and methods used to construct it. We then examine the generation of new understanding and proceed outward again to the growth and connections of the new understanding that can result from enhanced integration.

In particular the book examines these questions:

1. Why be concerned with integration in ecology?
2. What is understanding and how does it relate to integration?
3. What is theory and what are its parts? How is theory classified and how does it change?
4. What drives change in theory and hence change in understanding?
5. How, exactly, does change in understanding promote integration?
6. What scientific and social factors limit integration?
7. How does ecological understanding relate to the larger society?

In our discussion, several themes emerge. First, a broad view of theory is supported by modern philosophy and the history of science. This broad view links the empirical and conceptual approaches that are often considered to be separate. Second, an objective view of scientific understanding emerges that can accommodate the variety of seemingly disparate activities that scientists practice. Finally, we identify some large targets for integration in ecology.

This book is intended for anyone who has some background in ecology, beginning with advanced undergraduates. We do refer briefly to some ecological examples but must depend on other sources for the detail. To supply a large number of ecological examples here would obscure the broad picture of understanding and the use and structure of theory we wish to present. We hope the book will be useful and interesting to ecologists of all kinds. Of course, we hope it stimulates application of the general approach in a variety of ecological realms. Using the framework we present, ecologists should be able to assess the status of theory and understanding in their own topic areas.

We have received the good advice of a number of people on early essays and in discussions that advanced our progress on this book and clarified our thinking. We thank James H. Brown, Richard T. T. Forrnan, Marjorie Grene, Elizabeth A. Lloyd, Robert H. Peters, Peter W. Price, and Richard Waring for help along the way. We thank our colleagues at the Institute of Ecosystem Studies (IES) for providing a stimulating and open intellectual environment that made these explorations possible. We thank IES librarian Annette Frank for help in obtaining references and Sharon Okada for redrafting and improving some of our problem artwork. The financial support of the Mary Flagler Cary Charitable Trust, of the U.S. National Science Foundation for essentially "empirical" work (BSR 8918551; BSR 9107243) and for Research Experiences for Undergraduates (BBS-9101094), and of the Canadian Natural Sciences and Engineering Research Council has contributed to the instigation and completion of this book.

S. T. A. P. and C. G. J., Millbrook, New York
J. K., Hamilton, Ontario

PREFACE TO THE SECOND EDITION

We have often wondered why the second edition of a book needs a new preface and why the preface for the first edition remains intact. It always seemed like a quaint, librarian-like tradition. In case you are wondering the same thing, the goals, motivation, and organization of the book laid out in the preface of the first edition remain. If you are new to the book, be sure to read the original preface to the first edition. We are still trying to introduce the wider field of ecology to a philosophical view that can be helpful in integration and synthesis. In fact, we think that this need has only grown. As ecology embraces new areas, such as biocomplexity, guidance in the strategies and tactics for integration are, if anything, even more needed than they were a dozen years ago. Similarly, growth in the desire to link ecology with other disciplines has been shown to be increasingly important. So the perspectives and tools we bring together in this second edition are all the more important today than when we began the first edition.

The second edition is substantially revised and updated. While we retain many of the classic ecological examples we used in the first edition, we have updated the references underpinning these and have added many new examples. We have also reported on progress and new controversies that have arisen in the philosophical literature relevant to the topics we cover.

One major goal of this second edition is an attempt to increase the accessibility of the text. Some readers found the density of ideas per line made reading rather slow going. We have tried to reduce the idea density and to intersperse more examples to make reading and comprehension easier. We have also clarified passages that startled us with their stylistic complexity. The fact that they escaped our notice in the first edition was an unfortunate oversight. We have also taken this opportunity to add a number of illustrative diagrams and figures that reinforce or extend the message of the text. The use of text boxes has increased as well, while retaining the flow of the central text arguments, to permit their consideration and discussion as issues worth focusing on. Some of the boxes are intended to help readers recall key points.

This preface gives us the opportunity to add new acknowledgments beyond those in the first edition. S. T. A. P. thanks Dr. M. L. Cadenasso and a graduate discussion group of Dr. S. R. Carpenter at the University of Wisconsin for comments that improved the quality of the text. Dr. Cadenasso also helped put the bibliography together, which is much appreciated, and beyond that, her addition to our understanding of ecological frameworks has been profound. S. T. A. P. also thanks the owners and staff of the Armadillo Bar and Grill in Kingston, New York, for providing a welcoming venue for many productive Saturday afternoons of work on the manuscript.

J. K. thanks Dr. Martin Mahner and Greg Mikkelson for illuminating e-mail comments and Drs. B. Beisner and K. Cuddington for sharing earlier drafts of their book.

C. G. J. thanks the Institute of Ecosystem Studies for continuing support that has generated the opportunity for conceptual reflection.

This book is a contribution to the program of the Institute of Ecosystem Studies, with partial support from the Mary Flagler Cary Charitable Trust. Research supported by the National Science Foundation through the LTER program (DEB 0423476) and by the Andrew W. Mellon Foundation to the Mosaics Program at IES and the River/Savanna Boundaries Programme in South Africa generated examples used in this second edition.

S. T. A. P. and C. G. J., Millbrook, New York
J. K., Hamilton, Ontario

Part I

Advancing the Discipline and
Enhancing Applications

1

Integration in Ecology

"Science is a map of reality."
Raymo 1991:147

I. Overview

Two themes emerge from the diversity of ecological science, and these themes run throughout this book. First, there is a need for greater integration across the diverse discipline of ecology. Second, there is enhanced opportunity because new tools are available for integration. Paradoxically, the first theme, integration in ecology, arises from progress in the field's subdisciplines; the substance of ecology in specific subjects has advanced greatly over the past several decades. However, the progress of individual subdisciplines does have some negative consequences. Ecologists often debate whether the approach of one subdiscipline is better than that of another; or ecologists with training in different specialties approach the same question in seemingly contradictory ways. While different subdisciplines offer unique perspectives that can contribute to solving problems, much of the subdisciplinary debate within ecology is in fact damaging to progress. That damage can be repaired and prevented by integration. The resolution of divisive controversy is one benefit that ecology can gain from integration. As a consequence, integration can accelerate progress, advance understanding, and enhance the application of ecology.

The second, related theme is the tools needed for effective integration. These tools enhance the clear elaboration of sound scientific content. The basic concepts that are used in different ways across the breadth of ecological science require clarification. Finally, we must articulate the nature of the broad understanding that we seek. To appreciate how these tools are used, their relationship to novel philosophical insights about how science progresses is required.

Integration requires that we know what we understand, what we want to integrate, and how to achieve this. So this chapter lays out three goals of the book. First is an examination of what constitutes understanding and its components. Second is an evaluation of integration and how it might be accomplished in ecology. The third goal is an exploration of the relationships between ecological integration and its larger social contexts and constraints.

II. Ecological Advances and Diversity of Ecology

Ecology is a discipline of vast scope. It ranges from interest in how organisms affect the chemistry of the entire Earth, to how a particular physiological trait of an organism adapts to its local environment (Keller and Golley 2000, Likens 1992). It encompasses interest in the bacteria living in an Antarctic lake as well as interactions of people and environment in cities. It studies the effects of a sudden severe storm on a rocky shore and the changes in vegetation since the last ice age (Fig. 1.1). While the subject matter is vast, the range of motivations ecologists have is just as diverse. Some ecologists want to solve pressing environmental problems, some want to know how the integrated physical and biological world works, and yet others want to know how organisms interact with each other. The end result of such a wide array of subjects and motivations is a stunningly diverse discipline.

Ecology has made substantial progress in both basic understanding and application since its origins over one hundred years ago. There is much evidence of progress and intellectual growth. First, ecology has evolved a rich diversity of active subdisciplines, such as autecological, population, community, ecosystem, landscape, and global ecology (Fig. 1.2). New data, creative tests, and novel generalizations appear continually. Ecology contains a plethora of approaches encompassing the growth experiments of ecophysiology, the feeding trials of chemical ecology, the watershed experiments of biogeochemistry, the pattern analysis of macroecology, the elemental budgets of global geochemistry, and the models of ecological genetics, to name but a few. The number of ecological journals and publications has steadily increased, as have the diversity and membership of scientific societies that have an ecological basis. Such growth indicates focus on novel or neglected questions and the advent of new areas of research and new ways of thinking. Finally, the use of ecological information is increasing in such areas as environmental policy and management, conservation biology, restoration ecology, watershed management, and global environmental change (Orians 2005, Pace and Groffman 1998, Palmer et al. 2004, Turner et al. 2004). All of these are important and laudable developments.

Nevertheless, much of this growth has been focused within subdisciplines as reflected by the inauguration of increasingly specialized journals. According to an editorial in *Ecology Letters* in the first issue of 2003, ecological journals had increased by 60% in the previous decade. We present an illustrative analysis. Of 27 ecological journals listed in MedBioWorld beginning with the letters A and B, eleven started after the year 1990, and only six began before 1970. A good illustration of this balkanized diversity can be seen in the results of a poll of British Ecological Society that asked members to vote for the most important concepts in ecology (Table 1-1;

Figure 1.1   Illustrations of some of the variety of systems of interest to ecologists. Clockwise, from left: A. Sunbird and a giant *Lobelia* on Mt. Kilimanjaro, Kenya, illustrating physiological and vegetation studies of plants in cold environments, and plant-pollinator interactions. B. Fresh edge in primary lowland tropical rainforest, Costa Rica, representing studies of ecological boundaries, landscape ecology, invasion of exotics, and conservation biology. C. Rocky slope in the Negev Desert, Israel, representing patch dynamics and studies of natural disturbance and spatial heterogeneity and ecosystem function. D. Olifants River, riparian zone in middle ground, and upland savanna in background, Kruger National Park, South Africa, representing boundary dynamics, flood disturbance, fire, and plant-herbivore studies. E. Ephemeral pools on rocky beach, Jamaica, representing metacommunity and patch dynamics studies, and studies of physical stress. F. The Union Square neighborhood, Baltimore, Maryland, representing urban ecological studies and the integration of biophysical with social science research. G. Water "tank" at the base of a bromeliad plant, representing studies of food web dynamics, island biogeography, and metapopulation studies. H. Mown meadow in foreground and second growth deciduous forest at the Institute of Ecosystem Studies, New York State, representing studies of plant succession, animal dispersal processes, and boundary dynamics. These sites or similar ones are the subject of ecological studies. Rather than a comprehensive roster, this selection suggests the breadth of contrasts among the kinds of systems that ecologists study. Note the variety of spatial scales. In addition, local or regional human influence is a factor in many of these sites, and global changes in climate and atmospheric pollution loading affect them all. Photos A, E, and G by J. Kolasa. Photos B, C, D, F, and H by S. T. A. Pickett.

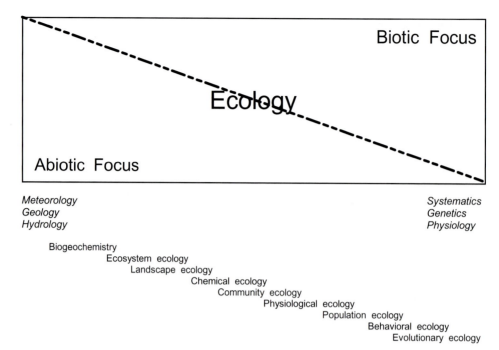

Biotic Focus

Ecology

Abiotic Focus

Meteorology   Systematics
Geology   Genetics
Hydrology   Physiology

            Biogeochemistry
                    Ecosystem ecology
                        Landscape ecology
                            Chemical ecology
                                Community ecology
                                    Physiological ecology
                                        Population ecology
                                            Behavioral ecology
                                                Evolutionary ecology

Figure 1.2   The topic gradient of ecology. Ecology is an extremely broad discipline that can be conceived of as linked
———   subdisciplines arrayed along a gradient that ranges from concern with strictly biological to concern with
                strictly physical phenomena. However, throughout most of the discipline, some mix of abiotic and biotic
                focus is necessary. Disciplines focused more narrowly on biotic or abiotic topics are shown in italics at
                the extremes of the gradient. Missing from this representation is the growing effort to connect ecology
                with social sciences and economics, which might be considered to represent a dimension orthogonal to
                the one shown. Reprinted with permission from Likens (1992).

Table 1-1 Ecology's Top 30: Ecological Concepts in Order of Their Rank in a Survey of the Membership of the British
Ecological Society

1. Ecosystem	16. Limiting factors
2. Succession	17. Carrying capacity
3. Energy flow	18. Maximum sustainable yield
4. Conservation of resources	19. Population cycles
5. Competition	20. Predator-prey interactions
6. Niche	21. Plant-herbivore interactions
7. Materials cycling	22. Island biogeography theory
8. Community	23. Bioaccumulation in food chains
9. Life-history strategies	24. Coevolution
10. Ecosystem fragility	25. Stochastic processes
11. Food webs	26. Natural disturbance
12. Ecological adaptation	27. Habitat restoration
13. Environmental heterogeneity	28. Managed nature reserve
14. Species diversity	29. Indicator organisms
15. Density-dependent regulation	30. Competition and conditions for species exclusion

Cherrett 1989). Many of the topics in the concept list are the focal concern of one or perhaps two of the many, diverse subdisciplines of ecology.

A. Consequences of Disciplinary Progress

At least three negative, but perhaps inevitable, consequences have arisen as a result of the rapid and admirable progress in ecology. First, with the development of subdisciplines, gaps in our understanding appear at the interfaces between those subdisciplines (Jones and Lawton 1995). For example, landscape ecology focuses on spatial heterogeneity in ecological systems, while ecosystem ecology focuses on fluxes of matter and energy within ecosystems. The gap that has arisen between these disciplines is the role of spatial heterogeneity in controlling ecosystem fluxes. Integration bridging this gap is currently being attempted (e.g., Cadenasso et al. 2003, Shachak and Pickett 1997). Gaps like this between disciplines beg to be filled and can even spur the creation of new subdisciplines. Indeed, ecology itself can be seen as an invention filling the gap between organismal physiology and biogeography (e.g., Schimper 1903; also see McIntosh 1985). In a world of increasing specialization, more attention has to be directed toward such gaps (Ziman 1985).

The second negative consequence of narrow progress is that disciplines tend to focus on specific scales or levels of organization (see Allen and Hoekstra 1992). For example, in the past the study of plant communities focused on fine scale structures and processes that could be found within a few hectares and that generated change on the scale of years to a few decades. This was traditionally considered a discrete level of organization suitable for focused ecological study. Plant population ecology represents a different level of organization, one that focuses on the demography of a single species in a circumscribed area. When these two levels were integrated in the 1970s and 1980s, understanding of how plant succession occurs was substantially advanced (e.g., Bazzaz 1996). The integration led to understanding succession as a process of interacting populations, its dependence on the differential, evolved allocation strategies as the fundamental basis for the interaction, the mix of early and late successional traits in many species, and the verification that succession was, as Gleason had proposed, at base, an individualistic process (Pickett 1976, Pickett and Cadenasso 2005).

Improvement of ecological understanding also results from integration across spatial or temporal scales. An example of this also exists in the study of succession. As knowledge of plant succession increased, ecologists became aware of the need for research bridging different scales. Successional studies now include influences beyond the obvious or convenient boundaries of a plant community to include historical events that took place before the succession started and processes that originated in adjacent or distant communities. In addition to advances within existing disciplines, changing the scale of focus has enhanced the establishment of new disciplines in ecology. Incorporating coarser scales of study aided the development of the field of landscape ecology. This entire discipline grew out of recognizing that spatially distant influences can generally affect local ecological systems. Organisms and materials can move from one patch to adjacent patches, such as from a field to a forest, resulting in new interactions in the original patch (Cadenasso and Pickett 2001, Cadenasso et al. 2004). Processes such as nutrient export from one ecosystem are the inputs to another ecosystem. It has became clear that the spatial arrangement of patches in nature could have an effect on the behavior of specific sites (Pickett 1998, Turner 1989). For example, the patches representing different successional states interact in the dynamics of rocky intertidal systems (Paine and Levin 1981). Another example of historical integration between disciplines and levels of organization is metapopulation ecology. Here, spatial processes are directly incorporated into population dynamics (Hanski and Gilpin 1997).

Third, as subdisciplines become rich in detail, they develop their own viewpoints, assumptions, definitions, lexicons, and methods. One negative result is that, in many cases, the same term can have very different meanings in different subdisciplines. For example, the terms "regulation," "function," "development," and "evolution" have quite different meanings in population, community, and ecosystem ecology. Since most ecologists have a focus within a subdiscipline, interrelating the viewpoints of different subdisciplines becomes increasingly difficult because the conceptual frameworks of the different areas diverge over time. For example, although physiological ecology and biogeography have common roots (MacArthur 1972, Schimper 1903), they barely intersect now.

B. Dichotomous Debate

Gaps between areas may result in unnecessary and unproductive debate. Dichotomous debate can also occur in the unoccupied territory that appears between hardened polar positions or hypotheses *within* a specialty. For example, debates over whether a community is a discrete unit in and of itself or is an assemblage of interacting populations have been persistent and sterile ones in ecology (cf. Parker 2004). Likewise, whether populations are internally or extrinsically regulated has been a thorny debate. This debate has often been cast in terms of the roles of density-dependent versus density-independent control (Fowler 1990) and between intrinsic versus extrinsic regulation of organism numbers. Progress toward resolution in debates about community structure or population regulation took place when features of the opposing arguments were appropriately combined (McIntosh 1985, Shipley and Keddy 1987). In reality, as is now well known, the determination of population size involves density-dependent and density-independent factors, intrinsic and extrinsic processes, as well as dynamic feedbacks between these processes (Krebs 1994). The concept of density vagueness is an attempt to incorporate the two poles of limiting effect (Strong 1984). A more recent and as yet unresolved debate involves the best approach to ecological experiments. Some have argued that microcosm studies are an efficient way of advancing insights (Drake et al. 1996), while others believe that only large-scale studies are relevant (Carpenter 1996). An effective reconciliation of the arguments is likely to benefit ecology and this reconciliation is likely to follow the path of other debates and find a resolution in synthesis.

Of course, not all debate is damaging. Debate that clarifies issues or forces decisions among two real choices is useful. However, a debate that fails to clarify the very issues that generate the debate is likely to be unproductive. Unfortunately, real dichotomous choice is rarely the case in ecology, since ecological systems are invariably contingent upon history and spatial context. We can apply the notion of integration within subdisciplines to other contemporary ecological debates. Numerous problems that have been characterized by persistent dichotomous debate have ultimately been shown to benefit from integration. We might cite the same need for integration in debates over the roles of competition and predation in the structure and dynamics of communities (Roughgarden and Diamond 1986), the role of local versus regional processes in community organization (Griffiths 1999, Srivastava 1999), the benefits of studying small or large ecological systems (Nixon 2001), the role of abiotic or biotic regulation of ecosystem fluxes (Bond and Chase 2002), and the role of null models (Hubbell 2001).

The relationship between stability of ecological systems—whether they are populations, communities, or ecosystems—and their complexity is another current example of unresolved ecological debate. While the underlying concepts refer to plausible relationships between functional stability and organizational complexity as reflected by diversity, diversity is often construed simply as species richness or its derivatives, and stability is often construed as a single measure of one function (e.g., aboveground primary production). Results reported from nature, field experiments, and natural and lab microcosms are often inconsistent and difficult to explain (Naeem 2002), with proponents sometimes unwilling to recognize contradictory results. Additional clarity about the assumptions, hypotheses, and taxonomic context may be necessary before the issue can be tackled successfully. For example, the different taxonomic context of plant versus animal systems or single trophic versus multitrophic systems may present different forms of diversity-stability relationships. Furthermore, construal of species richness as a measure of organizational complexity and singular estimates of process as measures of functional stability are particularly laden with assumptions. Nevertheless, if the debate about diversity and stability is resolved, it has great potential for helping integrate population dynamics, community interactions, conservation, ecosystem services, and many other areas of practical significance.

Unproductive debates may also result from tacit focus on different scales. For example, ecologically interesting structures are often labeled as a "boundary," an "edge," or an "ecotone," depending on the scale of focus and the research tradition. "Boundary" is the term of preference in physiological ecology and soil science where the focus is usually on the fine spatial scale. In community ecology, the term "edge" predominates, and discrete structure is emphasized. At the coarse scale, transitions between community types or vegetation groupings are usually labeled "ecotones." There are often assumptions about the nature of the persistence or intentionality of the causes of the transitions, but the predominant difference is one of scale (Fig. 1.3). However, discovering the basic, underlying idea of a structural or process gradient that can affect a different ecological process or structure leads to a unified concept of boundary that can apply at any spatial scale (Cadenasso et al. 2003). This recognition has begun to stimulate comparative and integrative studies across types of systems and research and modeling approaches that had previously been pursued independently (Belnap et al. 2003, Cadenasso et al. 2003).

All of these examples show that debate is often problematic in ecology, and it often arises from poorly articulated concepts or contrasts. The limitations of dichotomous debate make it reasonable to suppose that advances in understanding may be made by asking questions such as when, where, and why some processes are more important than others, rather than asking whether process A or B is the right solution. What determines the mix of forces in particular cases? Such questions require synthesis of existing data, as well as new types of studies. Ultimately, the process of integration should help resolve the dichotomies; afford greater powers of explanation, prediction, comparison, and generalization; and eventually lead to the disappearance of current rival "schools of thought" and their replacement by a unified approach. Of course, any new resolution may lead to a new generation of controversies that could, in their turn, also benefit from integration. Cycles of debate and integration may well run through in the history of ecology.

C. Ecological Paradigms

Because the concept of paradigm is, at least in a general way, familiar to most ecologists, we can use this important idea to show how ecology might be advanced by integration. A paradigm is the set of background assumptions that a discipline makes. Another way to summarize the idea of paradigm is that it is the worldview that the scientists in a discipline hold. Paradigms mold subject area, approaches, and modes of problem solving (Kuhn 1962). Criteria of observation — the perspectives taken, kinds of processes involved, and kinds of interactions included (Allen and Hoekstra 1992) — are often different between paradigms. This discussion will be developed more fully in Chapter 8.

The value of and need for integration become especially apparent if we consider the contrast between two of the largest paradigms in ecology (Jones and Lawton 1994). One represents population ecology, and the other represents ecosystem ecology in its traditional or commonly

Figure 1.3   A conceptual diagram of a model of boundary structure and function, and two contrasting kinds of ecological boundaries, to which the conceptual model applies. The boundary model shows the major components of any study of the structure and function of an ecological boundary. The components of the model are adjoining patches, including relevant architecture, composition, and processes, the structure of the boundary itself, the flux across the boundary indicated by the arrows, and how that flux might be altered by the boundary structure. Boundaries may facilitate, inhibit, or have no net effect on the flux across them. Modified from Cadenasso et al. (2003). A: A mangrove boundary between terrestrial and marine habitats. B: The riparian vegetation of the Shingwedzi River in the Kruger National Park, South Africa, acting as a boundary between upland and riverine habitats. The photograph was taken in the dry season when the flow in the river is much reduced. Photo A by J. Kolasa. Photo B by S. T. A. Pickett.

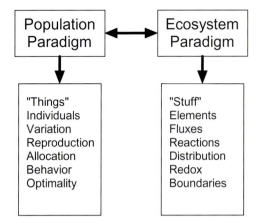

Figure 1.4 The contrast between the ecological paradigm focusing on entities or "things" and the paradigm focusing on fluxes or "stuff." The "things" paradigm is most frequently encountered in organismal or population ecology and is shown on the left. The "stuff" paradigm reflects the common view of the ecosystem ecology, shown on the right. Basically, the organismal paradigm is concerned with biological entities that can be differentiated and enumerated. The ecosystem paradigm is basically concerned with the controls on fluxes of materials and energy. Within each list appear some entities or processes that are of particular concern under each of these two major ecological worldviews.

perceived form. We will explore the broader manifestation of the ecosystem idea later. For now, the primary distinction between these paradigms is their focus on organisms on the one hand and on materials and energy flow on the other (Fig. 1.4). Such a distinction is clearly apparent in ecological textbooks that differ in their focus. The contrasting foci are reflected in the definitions of ecology found in textbooks that represent these two different paradigms (Box 1.1). We will expand on these differences next.

As representatives of the population paradigm, Begon et al. (1996) emphasized the individual organism, or populations and communities composed of individual organisms, as the basic units of study. The population ecology paradigm addresses patterns and causes of change in the distribution and abundance of organisms in space and time (Silvertown 1982). Abiotic factors are usually considered to be external forcing functions altering the dynamics of organisms and aggregations of organisms. In most cases, the ecosystem containing the population is taken as a given, for simplicity's sake. Thus, to answer the questions commonly posed by population ecologists, changes in material and energy flows have not usually been well connected to organismal dynamics (but see Cale 1988, Robertson 1991).

As an example of the ecosystem paradigm, Odum (1971) emphasized matter and energy flux in ecological systems, which links with his attention to physiology in the organismal realm of ecology. The common view of the ecosystem paradigm focuses on patterns of material and energy flow and the processes controlling them (e.g., Reiners 1986, Schlesinger 1991). The abiotic environment is explicitly included in the ecosystem and, of necessity, the complex embedded dynamics and heterogeneity of organisms are often "black-boxed" or treated as though the mechanism operated in a closed box, out of sight. Thus, the complex dynamics of heterogeneity and organisms are taken as constants (Cale 1988, Damuth 1987). This perspective is by no means static. Some ecosystem theorists believe that the cybernetic or closed-box approach is not a complete ecosystem perspective (e.g., O'Neill et al. 1986) and that the ecosystem concept itself needs major revision (O'Neill 2001). Debates about the nature of the fundamental concepts in a paradigm are not unique to ecosystem ecology — the population concept is continuously revised as well (e.g., Baguette and Stevens 2003).

BOX 1.1 Definitions of Ecology

Definitions are given, along with sources, for different ecological paradigms or viewpoints:

Ecosystem paradigm. The study of the structure and function of nature (Odum 1971).

Population paradigm. The study of the interactions that determine the distribution and abundance of organisms (Krebs 2001); the study of the natural environment, particularly the interrelationships between organisms and their surroundings (Ricklefs 1977).

Toward integration — organism centered. The scientific study of the processes influencing the distribution and abundance of organisms, the interactions among organisms, and the interactions between organisms and the transformation and flux of energy and matter (IES definition; see Likens 1992).

Toward integration — general. The study of ecological systems, and their relationship with each other and with their environment, where ecological system is defined as any natural or arbitrary unit at or above the organismal level of complexity.

Packing organismal characteristics away in an opaque box does not mean that they have no intrinsic interest or significance in other processes. Rather, this perspective is one approach to necessary simplification in ecosystem ecology, permissible because considerable progress toward the main goal of understanding ecosystem structure, function, and change can be achieved primarily by measuring material inputs, outputs, transformations, and pools. Fluxes are regulated by both the abiotic components and organisms in ecosystems. However, in most studies, organisms are aggregated on the basis of their role in processes of flux. The categories of producers, decomposers, N-fixers, and shredders are examples of the aggregations of organisms conceived in ecosystem studies. How the details of organism identities and dynamics influence fluxes is not usually central (but see Marks 1974, Vitousek 1989). Therefore, taking organism features and behaviors out of the box and examining the reciprocal effects of organisms and ecosystems form an active frontier for integration (Chapter 7).

Since the foci, scales, and criteria of observation in population and ecosystem ecology are usually so different (Fig. 1.4), their paradigms and constituent assumptions, approaches, and lexicons are also different. These differences are rational and helpful for focusing research. No doubt major advances in ecological understanding will continue to be made in each of these subdisciplines. However, in the broadest sense, ecology is "the scientific study of the processes influencing the distribution and abundance of organisms, the interaction among organisms, and the interaction between organisms and the transformation and flux of energy and matter" (Likens 1992). From the perspective of this broad definition emphasizing both organismal and systems properties, it is clear that integration of the population and ecosystem subdisciplines is a goal for advancing ecological understanding. Such a goal can be achieved by forging links across these subdisciplines and by focusing on ecological issues and critical questions that lie at the intersection of the subdisciplines (Jones and Lawton 1994) and, therefore, cannot be addressed by either of the two paradigms alone but require input from both paradigms.

Another possible avenue to integration, which we will explore more extensively in Chapter 4, is through increased generality of the basic framework, including its lexicon and classes of relationships (Box 1.1). This route has successfully been taken in some areas of physics (e.g., unification of electromagnetic phenomena) and may be useful in ecology. It may be that taking "system" in a broad and inclusive sense, with clear definition of scale, scope, and phenomena included in an ecological system, will be helpful to integration.

D. Integration and the Role of Ecology in Society

We have proposed that integration can play a large role in advancing ecological understanding among and within subdisciplines. These benefits accrue to the science itself and do not necessarily result in any social benefit. However, clear benefits can lie at the intersection of ecology and society (Odum 1996), and these may also profit from increased integration. For example, dealing effectively with human-accelerated environmental change (Likens 1991) requires integrating many topic areas, scales, and levels. Indeed, one recent realization is that ecology must integrate more effectively with social and physical sciences, especially on regional scales, if it is to contribute to the solution of important environmental problems (Grimm et al. 2000, Groffman and Likens 1994, McDonnell and Pickett 1993). As a result of effective internal integration and integration with other disciplines, ecology can become increasingly perceived as advancing rapidly and contributing to the solution of societal problems. Scientists in other disciplines and society as a whole need to see this face of ecology. Since ecology has many interfaces with other scientific disciplines such as geochemistry, climatology, genetics, and biochemistry (see Fig. 1.2), increased integration is likely to attract more interdisciplinary collaborations (Berry 1989). Increased representation by ecologists in national and international interdisciplinary scientific forums might be another consequence. As managers, politicians, and the general public gain increased awareness of and use of ecological findings, it becomes easier to earn their support (Pickett 2003). A further advantage of integration within the discipline is the increased communication between ecologists, bolstering our ability to identify and present new issues worthy of support and intellectual effort. Ultimately, how to integrate with managers and policy makers is also an important synthetic strategy for ecology (Rogers 1997, Shrader-Frechette and McCoy 1993).

III. Progress via Integration

We have introduced the opportunities for integration suggested by several features of ecology. One is the gaps that exist between specialties and polar alternatives. A second is the bridging of different spatial and temporal scales. The third is the unification of two major ecological paradigms, as illustrated by the ecosystem and population perspectives. Now we are in a position to examine some of the circumstances that will promote the success of integration (Pickett 1999). Although we will examine the component parts of integration more rigorously later (Chapter 7), we introduce a working definition now so that considering how to promote the process has a clear context. Simply, integration is the growth of connections among existing data, perspectives, approaches, models, or theories that are apparently disparate. In other words, integration is a linkage between different conceptual views of the world or different kinds of data.

Integration among and within subdisciplines is clearly a desirable goal that can advance the discipline of ecology as a whole without impairing progress within subdisciplines. However, achieving this goal is not easy. Here we will preview some of the tools needed for integration and set the stage for more complete analysis in later chapters. Four classes of tools are required: (1) domain, (2) conceptual clarification, (3) scale and level, and (4) methods.

A. Domain

Integration is difficult to attain because it requires an adequate state of advancement to exist in two or more subdisciplines. The conceptual framework of each subdiscipline must be well formulated, and a new framework that links both must be developed. This requires careful attention to the basic focus, or domain, involved. The domain states the phenomenon and scales of interest. The scale includes both the extent, or largest range of observations that will be used and the grain, or resolution of the observations. More will be said about domains in Chapters 2 and 3. Without clear statement of the phenomenon and scale of interest, what is to be joined and what the potential points of contact might be will not be clear enough to proceed. For example, to promote the unification of ecosystem and population approaches requires an understanding of what each paradigm usually contains.

B. Conceptual Clarification

Some of the problems identified in examining domain require conceptual clarification. Concepts in ecology are remarkably complex and multidimensional entities. There are two large considerations that can help ecologists use concepts more clearly. First is to recognize that concepts have three dimensions: a core definition, a modeling strategy to apply the concept to real or simulated situations, and a set of metaphorical implications that are used in informal, creative, and cross-disciplinary communication (Pickett and Cadenasso 2002). Second is to recognize that concepts often can refer to both process and outcome. We address these two major aspects of concepts in turn.

Concepts in ecology, indeed in science in general, have three dimensions. For ease of discussion, we refer to these dimensions as (1) meaning, (2) model, and (3) metaphor (Pickett and Cadenasso 2002). The *meaning* of a concept is the core or most fundamental definition it has. Such a core definition would be general and would underwrite a broad domain. For example, the concept of ecosystem refers at its core, to a biotic complex and an abiotic complex in a specified area (Likens 1992, Tansley 1935). The use of the root "system" in the term "ecosystem" confirms that the core definition also includes the interaction between the abiotic and biotic complexes in that area. Tansley was very clever in stating the core definition so broadly. It can apply to any unit of the Earth and can be expressed on any temporal or spatial scale. If the ecosystem idea is taken as an exemplary case, then we should expect all core definitions of the most important ecological concepts to be general, scale independent, and broadly applicable.

To use the core definition of an ecological concept, the second dimension of concepts is required. This additional dimension is the application of a concept in the form of explicit models. In other words, a general concept requires a model to make it fit the world. The "world" can mean a jar on a laboratory bench, a field experiment, a simulation algorithm, or a study area outside. Models take the basic, general idea embodied in a concept and indicate what specific place or kind of place it refers to, what the spatial and temporal scales are, and what structures and interactions are expected to hold there. A single concept may result in many models. For example, the concept of ecosystem can apply to a laboratory aquarium with measured inputs of energy and minerals, and the system may exist for the duration of an experiment. The concept can also be applied to the entire planetary ecosystem, as when the biosphere of the Earth is modeled. Or the concept may be operationalized as a system of equations that generate energetic, material, or biodiversity outputs. In other words, there is an almost inexhaustible range of models that can be generated from the core concept of the ecosystem. Each model specifies the details of the real or imagined world that give structure to the abiotic and biotic complexes and the linkages and feedbacks between them.

Table 1-2 Ecological Concepts Expressing Net Effects of Interactions and Transformation
These are logical types only, and the intensity of the interaction is ignored in this classification. Succession as used here is the classical *facilitation* interpretation, in which pioneering species alter the environment to their detriment, while the changes favor later successional species.

Net effects						
Interaction	+/−	+/+	+/0	0/−	−/−	−/+
Competition	X				X	
Predation	X					
Parasitism	X					
Commensalism			X			
Ammensalism				X		
Symbiosis		X				
Succession						X

The third dimension of concepts is their metaphorical connotations. Such connotations may be the impressions that members of the public or practitioners of another discipline think of when they encounter an ecological term. For example, the "ecosystem" may connote connectedness, stability, and diversity in the minds of citizens who know the term. Note that none of these conditions, except connectedness, is required by the definition. Note further that even in the case of connectedness, the definition is silent on the density, strength, or effectiveness of connections. These are empirical matters to be determined for specific models or specific instances of ecosystems. However, the connotations are powerful and unavoidable baggage that the concept carries into any discourse that is general and independent of a specific model and case. Metaphors are useful in initiating discussion between specialists in different disciplines or subdisciplines, and between scientists and members of the public.

So far we have seen that conceptual clarification relates to the three dimensions of any concept: (1) identifying the core, most generalizable definition of a concept; (2) understanding how the concept is operationalized in models for specific places or circumstances; and (3) appreciating what the metaphorical connotations of the concept are. These dimensions can be summarized as meaning, model, and metaphor or, alternatively, as definition, specification, and imagination. Where one is working in this three-dimensional mental space is crucial to the clear use of concepts.

An important additional issue exists in ecological concepts that refer to processes. The word denoting such a concept may refer to either the process itself or to the outcome of the process. In other words, a conceptual term in ecology can sometimes refer to an action or mechanism and at other times can refer to an outcome of that action. The problem of a single concept referring to an underlying process and the outcome of that process is common in ecology and is often the source of confusion and dichotomous debate. We can exemplify this problem using the concept of competition. The term "competition" sometimes refers to (1) the *process* of joint use of a limiting resource (e.g., Keddy 1989), whereas at other times it refers to (2) the *outcome* (e.g., Odum 1971) — that is, the success of one organism over another or the reciprocal effects of two populations on one another (Table 1-2). In the first case, the process of two organisms jointly using a limiting resource results in a decrease in abundance of the inferior competitor. The focus is on the process, and the outcome is understood in light of knowing that the resource is actually limiting, that one organism is more efficient in using or effective in obtaining the resource than another, and that the reduction in one species is the result of the interaction of the two species mediated through that common resource. Knowing the difference between the process and the outcome is important, because an alternative outcome of joint use of resources may occur. There may be no decrease in abundance of the focal species due to the intervention of some other process, such as the emigration of the more successful competitor or its suppression by disturbance. In other cases, an organism may depress the abundance of another species via indirect effects, such as alteration of habitat or reducing the positive effect of a third party, which is termed "apparent competition" (Holt 1977). Although presenting a detailed solution to this problem of conceptual clarity in competition is beyond the scope of this book, in essence, the solution lies in recognizing the difference between process and product (e.g., Brandon 1990). Focus can be on the processes — that is, the mechanisms or specific interactions — or on the products — that is, the outcomes or net effects of competition. Once one recognizes the sometimes obscure difference between process and product, it becomes clear when and where similar or different processes lead to similar or dissimilar outcomes. Therefore, although only three possible outcomes or net effects on abundance can result from interaction of two species (increase, decrease, or no net change), many processes or mechanisms may be responsible for these effects. Furthermore, this distinction between process and product invokes recognition of the importance of the rest of the system in which the interacting entities are embedded. While the process of competition can be understood by studying two players and their resource in isolation, the outcome depends on both this process *and* the degree to which these entities are connected to other components of the system. The examples given earlier where outcome and process are not the same — an intervening process and "apparent competition" — illustrate this point.

Conceptual clarification often triggers new questions. In the preceding example, the question "Is competition important?" might be replaced by several questions: When, where, how, and why does joint use of a limiting resource result in decreases in abundance? What factors in the system in which embedded players interact are responsible for there being no net change in abundance despite the fact that species share a common limiting resource? Or, is intensive competition a stabilizing or destabilizing factor, depending on other circumstances? Thus, explicit specification of the concepts and domain underlying a research question can improve chances for comparison, generalization, or integration among cases.

C. Scale and Level

Because integration often crosses organizational levels, careful attention to scale is critical. Thus, in our example of competition, we might have to pay particular attention to population, community, and ecosystem "levels of organization" together. Ironically, even these levels exemplify confusion within the common vocabulary of ecology as classes of objects cannot constitute levels. Levels are often viewed as different degrees of aggregation of presumed basic units. If we take species as a basic unit, we may observe that two species, A and B, share a common resource and that the density of species B is negatively correlated with the abundance of species A. This could be because species A uses the resource more effectively than species B, which requires understanding of the autecological mechanisms of resource use. This explanation is constructed within one level of aggregation, that of the species population. However, the outcome could also be because species A interacts with the resource in such a way that the resource quality is altered and species B is depressed, requiring an understanding of the effects of A on the resource. This explanation occupies two levels of aggregation, that of the species and of the resource. Alternatively, it could be because another species, C, interacts with the resource, requiring community-level information. Also, it could be that the actions of species A on some other resource alter the capacity of species B to use the shared resource, requiring an understanding of ecosystem processes. Of course, several such "alternatives" can act together. Thus, answering our question would force us to move up and down organizational "levels" (i.e., be simultaneously reductionistic and holistic; Thornley 1980). Later, we will introduce a more refined look at levels of organization in ecology, one that leads to recognizing different descriptions of nature and their criteria (Allen and Hoekstra 1992), as well as their implications for formulating fundamental questions of ecology (Chapter 6) or resolving contrasts such as one between single versus multiple causality (Chapter 8).

D. Methods

The requirement that questions be formulated with an awareness of scale, the existence of models, and multidimensional concepts have important consequences for deciding on an appropriate methodology. Experiments may be more effective for some aspects (e.g., to investigate resource use efficiency or resource modification by species A at a particular site), whereas pattern analysis with its requirement for extensive spatial and temporal scales may be appropriate for other aspects (e.g., to correlate species A, B, and C or to define fluctuations in the other resources). Prediction may or may not be immediately feasible. For example, if A and B interact solely via differences in resource use efficiency, outcomes may be immediately predictable (cf. Tilman 1982). On the other hand, if interactions with other species or resources are involved, reliable predictions may not be achieved until the other interactions have been discovered, investigated, and found to have consistent effect on the focal process by, for example, being sufficiently repeatable. An excellent example of such a situation is a study on trophic cascades regulating salt marsh primary production (Silliman and Bertness 2002). These authors found that an overharvesting of crabs led to release of snails from predation and subsequent overgrazing of *Spartina*. Before their study, this important trophic effect was unsuspected and unrecognized.

IV. Integration, Understanding, and Theory

We have proposed that advances in ecology might be most effectively increased by enhanced integration within and among ecological specialties. We have adopted the definition of integration as establishment and growth of connections between apparently disparate areas, be they empirical, conceptual, or paradigmatic. The history of science shows that scientific progress is based on a combination of advancement within subdisciplines, the creation of new subdisciplines, and, most importantly, integration across subdisciplines (Cohen 1985). In other words, the most spectacular advances generally are born by interdisciplinary integration. Integration of such grand scope is not likely to proceed rapidly if it depends on empirical advances alone. Integration can only proceed if its context is clear and workable. This is difficult because integration may take scientists into conceptual territory for which no framework or map exists. The lack of frameworks in areas to be integrated may be a pressing problem if it is true that the unknown, rather than the imperfectly known, is the most profitable focus for progress in ecology (Slobodkin 1985). Conceptual framing may help point out the difference between the unknown and the imperfectly known. The imperfectly known will usually exist within a well-developed and well-articulated framework. Conceptual framing can help identify ideas, data, models, or paradigms between which connections can be established or enhanced. The unknown is the territory between the existing data, models, or paradigms. The open territory is only identified by mapping the existing knowledge or perspectives in the same frame to see what is missing. Without a good map — that is, without a framework — it is difficult to identify the unknown.

What serves as the framework for integration? Integration enhances understanding, and understanding requires theory. Therefore, one of the goals of this book is to present a broad-based view of theory as a key component of the improved understanding that integration represents. We will present the details of the nature of understanding and its dependence on theory later (Chapters 2 and 3). Before we present the details of how theory and integration are related and how understanding is used to evaluate integration, however, we must expose the key motivation for linking integration, understanding, and theory.

A philosophical perspective provides a great deal of information about these linkages, but the classical philosophy of science, which is the one that ecologists appear to know most about, presents a narrow, inappropriate, and outmoded view of theory and of the methods and development of science. Because the classical philosophy has been normative, telling scientists how to practice their craft, a narrow and problematic philosophy of science may, in part, limit the advance of ecology. We therefore present a brief overview of some highlights of contemporary philosophy of science that support a more useful view of theory for ecologists. These insights from philosophy can also be of great value for many concerns in ecology other than integration. Because we are practicing ecologists and not philosophers, however, we highlight only those ideas that seem to be readily applicable to ecology. Philosophers would undoubtedly be more concerned with the completeness and depth of the overview, exposing internal controversy about the philosophical points or the refinement of the philosophical ideas than we, as practicing scientists, can afford to be. Before we put the new philosophical information about theory to work in ecology, we put the philosophical novelty into perspective by first characterizing the old views of theory.

A. Points from the Classical Philosophy of Science Relevant to Ecological Integration

Even a brief overview of the vast discipline of the philosophy of science could start from many places. However, because, in our experience, the tenets of Karl Popper's philosophy are likely to be the most familiar to ecologists, we begin there. We find that ecologists are most aware of Popper's emphasis on falsifiability. That doctrine states that for a scientific *statement* to be meaningful, it must be falsifiable (Popper 1968). Falsifiability requires that an empirical comparison of the statement with the natural world be capable of exposing any incongruity between the statement and the relevant patterns or processes in nature. Such incongruity would indicate the statement to be false.

Falsifiability, on the surface, appears to be an easy criterion. A statement about the spatial relationship of the solar disk to the Earth's horizon is falsifiable, whereas a statement that the sun rises because the spirits have been appeased is not. The second statement is not falsifiable for several reasons. What a spirit is in this case and how it might have anything to do with the Earth and the sun is unclear. Popper and his contemporaries were concerned with differentiating just such contrasting sorts of statements. They were worried that the claims of Marxism, for the historical inevitability of proletarian revolutions, or Freudian psychology were being erroneously accepted as scientific. However, Popper's doctrine of falsifiability as a demarcation between science and nonscience indicts statements that are not so patently silly to scientists as the example about spirits and sunrise (Box 1.2). A statement that communities in resource-rich environments are probably controlled by competition for resources is also not falsifiable, but this is a probabilistic statement whose degree of explanatory power can be evaluated statistically by examining a large number of cases that represent a wide range of levels of resources and richness. Another example of a strictly nonfalsifiable but scientifically useful statement is this: "If organism a is better adapted than b in environment E, then a will probably have greater reproductive success than b" (Brandon 1990). In this case, the term "probably" is used in a rigorous, statistical sense. This second statement is an interpretation of the principle of natural selection, a schematic statement that embodies a generalizable mechanism of evolution. What is testable in the case of

BOX 1.2 Philosophical Insights

Philosophers have weighed in on important issues of methodology that are relevant to ecology. Here are some important highlights that can help ecologists clarify their thinking on these matters.

Demarcation

"There are several ways to characterize the logical empiricist project in the philosophy of science. Perhaps the best is to see logical empiricists as addressing the problem of demarcation — the problem of distinguishing between science and nonscience." (Boyd 1991:5)

Falsificationism

"Popper, whose views on this matter were published in German in 1934 but only became available in English translation in 1959 shared with early logical empiricists the conception that it was the testability of scientific theories that distinguished them from unscientific theories. He rejected, however, the verificationist conception that the possibility of confirmation or disconfirmation is the mark of the scientific. Instead, he was led by reflection on the fact that no inductive inferences are deductively valid to the conclusion that, strictly speaking, observations never confirm any general theories, but only refute or fail to refute them.

In consequence, Popper proposed a variation on the empiricist solution to the demarcation problem: a theory is potentially a scientific theory if and only if there are possible observations that would falsify (refute) it. The special role of auxiliary hypotheses in theory testing poses a challenge to this account of demarcation just as it does to traditional verificationism; but despite these technical difficulties, it remains deeply influential outside professional philosophical circles." (Boyd 1991:11)

Covering-Law Model

"It lays down criteria that a satisfactory explanation should meet, and it explains why we should value theories with explanatory power. But unfortunately matters are not so simple. In fact, the covering-law model faces a number of serious problems, widely (though by no means universally) regarded by philosophers of science as decisive." (Gasper 1991a:292)

Reductionism

"The existence of redundant causal factors . . . raises serious doubts that higher-level explanations in biology, psychology, economics, and sociology ultimately will be reduced to, or eliminated by, microexplanations." (Trout 1991:391)

Focus on Physics

"This preoccupation with physics is now commonly agreed to have had a distorting effect on the philosophy of science. The tendency has been to assume that certain features of physical theories, such as their tractability to mathematical axiomatization, are characteristic of scientific theories in general. To the extent that theories in other areas have not shared these features, it has been assumed that they are incomplete or deficient and that they need to be developed to fit the model derived from physics.

The "received view" of scientific theories articulated by logical empiricists from the 1920s to the 1950s is beset by serious internal difficulties. In other words, the dominant model of scientific theorizing seems inadequate even as a characterization of its central domain. In recent years it has become increasingly apparent that this model is even less appropriate for scientific fields other than physics." (Gasper 1991b:545)

natural selection is whether the scheme applies to a certain case or certain organisms. A test, therefore, involves determining the existence of organisms and environments that satisfy the general scheme. The lack of falsifiability of the general principle is only a problem under the assumptions and limitations of the philosophy that spawned the doctrine of falsifiability (Brandon 1990). Probabilistic statements are legitimate and powerful tools for ecology and for other sciences as well.

Popper was reacting to flaws in an influential school of philosophy called logical positivism, which developed in Vienna in the early 1900s (Boyd 1991). Logical positivism took on the problem of *verification* of theories as one of its main tasks. Logical positivists considered the hallmark of scientific statements to be that they were verifiable. Popper, however, thought that logical positivism did not successfully differentiate science from pseudoscience because verification, which was based on inductive reasoning, could not be deductively defended. Popper and his colleagues held falsifiability in such high regard because, under the rules of logic, no statement can be empirically proven. Inductive proof is contingent because an exception may lie just around the corner. Note that "logic" means here the specific field that studies the abstract structure of arguments. This mismatch between empirical evaluation and the narrow logic of proof was the stimulus for Popper. Therefore, Popper developed his "criterion of falsifiability" to divide scientific statements from unscientific statements.

The criterion of falsifiability was considered for a long time to be wildly successful, but many philosophers now conclude that in its logical elegance and simplicity, it leaves out many of the activities and products that are, in fact, legitimately part of science (Boyd et al. 1991, Grene 1984, Hacking 1983, Thompson 1989). For several reasons, both logical positivism and the Popperian alternative drew an incomplete, if seemingly compelling, picture of science. Presenting some of the reasons that Popper's picture was incomplete can help show how ecology is ill served by both the old philosophy of logical positivism and the contemporary holdovers of its antagonists. The old philosophy seems particularly unsuited to addressing integration in ecology. Here are the main reasons why we should avoid the old philosophy and its problems.

Physics was considered the exemplar of all science under the old philosophy of science. Logical positivists were rightly impressed with the success of classical physics and assumed that all science should and would become like classical Newtonian physics. However, accepting physics as the epitome of science had several negative consequences (Box 1.2).

The first problematic effect of taking classical physics as the exemplar of all science arises from the nature of theories in that discipline. Those theories may not be appropriate models for all useful kinds of theory. The theory of classical mechanics can be presented as a unified series of statements. The Newtonian laws of motion are composed of three statements that, as well as specifying the state of the system and inputs into it, can be used to predict the motion of bodies. The laws are said to be "covering laws." In the case of Newtonian physics, the statements or laws are mathematical in form. Philosophers took the laws of motion as their model for all scientific theories. This assumption resulted in what is now known as the "statement view" of theories. Under this view, a scientific theory is considered to be a series of statements (Carnap 1966, Hull 1974, Miller 1987, Rappoport 1978, Thompson 1989).

The laws of classical physics were also considered to be literally universal, applying to all bodies at any time and place in the universe. Any exception to a supposed literally universal statement meant that such a law was not universal. A supposedly universal law that was not literally universal was false. For this reason, falsificationism seemed a reasonable doctrine in the context of logical positivism. The laws of other disciplines, being necessarily more specialized, are not literally universal, although they may apply to a stated "universe of discourse." They apply to organisms or molecules as universes of discourse or domains. More importantly, however, the laws or general statements of other disciplines, including ecology, are often existential or probabilistic. Existential statements posit the existence of some entity or process that has certain characteristics or behaviors that would satisfy general laws. Such existential laws deal with variation and central tendency in ecological systems. This is an important feature of laws because ecological systems differ from one another as a result of their particular histories, genetic variation, or evolutionary or developmental state. Ecological laws must therefore take into account that they are intended to apply to evolving or historically contingent parts of the universe. Furthermore, ecological systems can be governed by multiple causes. Many different causes can interact to affect the outcome of a particular process of interest. Probabilistic statements account for uncertainty or variability. Therefore, ecological laws or statements may legitimately say, "The more isolated communities are, the more they are controlled by limited dispersal of organisms"; "Most crows are black"; "Nitrogen flux from river systems depends on human population density in the catchment" (Peierls et al. 1991). A level of probability could be generated for each of these statements (e.g., a p value and an R^2 for the N example; Peierls et al. 1991). Of course, in each of these cases, there may be other causes at work, and for the different instances, the mixture of factors now and in the past may be different. Communities may be controlled by processes in addition to dispersal. The color of crows may, in the case of a mutation, differ from black. And river nitrogen loading may depend to some extent on the quality of the wastewater infrastructure as well as the density of human population in a particular site. These probabilistic, multicausal sorts of laws are not unique to ecology. It is significant that the statement view of theory, based as it is on deterministic universal statements, does not work well in many contemporary areas of physics, where quantum uncertainty and complexity reign (Box 1.2; Cooper 2003, Gasper 1991b, Thompson 1989).

The statement view was also encouraged by the success, generality, and apparent practical power of Euclidian geometry, which is a root of the philosophical tradition that led to philosophy of science (Danto 1989). Following the geometric model, scientific theories could be likened to the elegant proofs that geometry provided. Of course, geometry has no empirical content. There are no field or laboratory data, but only ideal axioms as the starting point. However, in natural science, an empirical content replaced the ideal axiomatic origins of geometry. The fact that scientific theories were conceived of as sets of statements or laws, which governed the behavior of a system, prompted Popper and the logical positivists to struggle with a "criterion of demarcation" between statements that constituted science and those that pertained to other ways of dealing with the natural world. If science, philosophy, religion, poetry, and so many of the other human pursuits could be cast as series of statements, then demarcating science from nonscience became a pressing problem. Into that battle Popper entered with the powerful weapon of falsifiability, now seen as problematic (Amsterdamski 1975, Boyd 1991, O'Hear 1990, Putnam 1975, Thompson 1989).

The second problem with taking physics as the only legitimate model of science is that it narrowly determines how cause is conceptualized in other disciplines. The nature of causality in

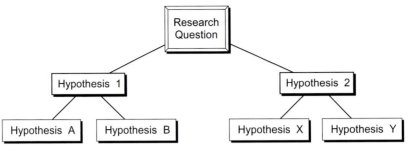

Figure 1.5   Pairwise alternative hypothesis testing. Alternative hypotheses aimed at answering a question are often paired as alternatives, implying a qualitative answer — all or none, yes or no — and discarding of one hypothesis from each branch. This approach is closely associated with the prescriptions for strong inference (Platt 1964). The guiding research question heads the hierarchy, which then presents two alternative hypotheses (1 and 2). Once one of those hypotheses is rejected, lower level, more detailed hypotheses about contributing causes or constraints are tested (A and B or X and Y).

classical physics is relatively simple. For example, collision of particles directly and unambiguously imparts momentum, and any alteration of trajectories or velocities relates directly to such collisions. Also, action at a distance via gravity follows clear, albeit exponential, relationships. These ideal behaviors from physics, cast as laws, are taken as evidence that single causes could be successfully invoked to explain events. However, complexity rears its head even in physics. Two body collisions are the models on which the success of classical mechanical laws resides. But collisions involving even a mere three bodies are notably unpredictable and chaotic in the strict sense. Chaotic phenomena, those caused by deterministic forces and subtle differences in initial conditions, or self-organized critical phenomena also are more difficult to predict or are inherently unpredictable in any detail. So causality has been recognized to be complex even in physics.

Direct, simple causality led some scientist-philosophers to propose "strong inference" as the ideal mode of scientific problem solving (Platt 1964). Because causes were simple and separable, they could be discriminated by a crucial experiment or observation. Under the procedure of strong inference, a tree of logically alternative hypotheses could be constructed, and the right cause could be disentangled from all the "wrong" or ineffective causes by a sequence of tests pitting the hypotheses against one another in a pairwise fashion (Fig. 1.5). This strategy does, in fact, work well when causes are separable and can be discriminated qualitatively. A qualitative discrimination means that whether a cause operates or not can be answered simply yes or no. Falsification is well served by the tactic of strong inference. In microbiology or clinical practice, a specialized version of strong inference is enshrined in Koch's postulates for isolating the cause of a disease (Box 1.3). Koch's postulates work well when the disease is caused by a single agent that can be isolated from other microbes that operate in the system and when other causes do not affect the action of the disease agent in unpredictable or nonlinear ways. These postulates have little to contribute to the solution of diseases with complex causes, including environmental agency and genetic propensity.

Another important outcome of using classical physics as the example by which all science could be understood was the highly developed state of physical theory near the turn of the 20th century (Shapere 1974, Hacking 1983). The laws of physics at the time that the philosophy of logical positivism was being developed constituted a well-elaborated theory. The laws and assumptions were clear, and the derivation of the laws from assumptions and from one another was known. In addition, classical physics had by then been highly confirmed by two centuries

---

BOX 1.3 Koch's Postulates (after Brock 1966)

Assume that the agent and the host are available in pure culture separate from one another.

The presumed agent must be shown to be present in active form when the change is taking place.

The agent must be shown to be present in larger numbers where the change is taking place than where the transformation is absent.

The agent must be isolated and grown in pure culture.

The agent when inoculated into a sterile host should produce the characteristic change.

---

of test, explanatory success, and practical application. Philosophers paid little attention to the origin and development of new disciplines and nascent theories but used older well-established disciplines and theories as their guides (Rosenberg 1985, Darden 1991). Diffusion of these views into ecology led some ecologists to pessimism about the prospects of ecology conducting research that was up to the physics standard or of discovering truly general laws (e.g., Roughgarden 1984, Lawton 1999).

The recommendations and advice for the conduct of science that emerged from logical positivism with its foundation in the established science of classical physics had no guarantee of success in developing, rapidly changing disciplines. The principal ecological professional organizations in the English-speaking world (the Ecological Society of America and the British Ecological Society) are younger than logical positivism.

Another old philosophical viewpoint — induction — sometimes influences how ecologists view theory. Following the process of strict induction, observations are assembled into tentative generalizations, termed hypotheses. With the accumulation of additional observations, hypotheses are converted by induction to theories. Finally, theories are confirmed as laws by successful application and extension to other observations (O'Hear 1990). This caricature can be labeled "the inductive chain," which shares with logical positivism the view that theory is a collection of statements. The problem of how a law can be confirmed by induction was one of the issues that motivated the logical positivists. They rightly noted that a generalization could not be *logically* confirmed strictly by the accumulation of data that supported it, because there was no guarantee that other unknown situations would come to light that would not support the generalization. The logical positivists were concerned with the logic of the scientific process and, on a strictly logical basis, confirmation by the amassing of positive cases was not defensible. Rather, confirmation, or the problem of induction, turned out to be an empirical matter (Miller 1987, O'Hear 1990). The laws of logic, though useful in science, are not the whole story. The empirical evidence and the multiple ways to approach it weigh heavily in the contemporary scientific process. More will be said about this later.

The overview presented here cites some of the critical aspects of the philosophy of science that, when imported into ecology as methodological assumptions or commandments, can influence integration. The dominant philosophy of science throughout the 20th century has been normative — that is, it has stated how science *should* be done. Logical positivism was generated in response to problems that rarely impede practicing scientists. Logical positivism was also based on a particular science, classical physics, that seems to be characterized by simple causality and unified, established theories that can be cast as statements. As a result, that classical philosophy has severe limits when applied to other scientific disciplines (Thompson 1989). Carte blanche application of attractive aspects of logical positivism and the statement view of theory and its derivatives such as falsifiability can create problems outside the realm in which they were formulated (Bauer 1992, Mayr 1982, 1988, Thompson 1989). Although certain aspects of logical positivism may be appropriate for some uses, newer philosophy of science (Grene 1984) has much more to offer ecology.

B. Aspects of the Emerging Philosophy of Science

The new philosophy of science is based on broad explorations of the history of science, rather than focusing on ideal practice. It also looks at other disciplines besides classical physics (Grene 1984, Thompson 1989, Boyd et al. 1991, Gasper 1991b). The new philosophy of science is not based on the statement view of theory but recognizes that theory can take many forms and that links between theoretical constructs and facts must be its integral components (Mahner and Bunge 1997). Statements still maintain a place in theory, but many other kinds of components of theory exist as well (Chapter 3).

The new philosophy of science maintains a pluralistic view of science, rather than assuming that there is one way to practice science. It applies to sciences that require various causal structures, not only the simple direct causality of classical mechanics (Salmon 1984). Moreover, it admits many tactics other than falsification for conducting scientific research (Hill 1985, May and Seger 1986). Also, the new philosophy has explored sciences that emphasize explanation and whose objects of study are strongly influenced by their histories (Lloyd 1988, Miller 1987).

One example of the pluralism in philosophy is the work of van Fraassen (1980), empirical constructivism Box 1.4.

Second, a philosophical perspective of potential appeal to practicing scientists is known as scientific realism. We leave these topics without further explanation because an exposition of various philosophies of science is not our goal. We mention this philosophy of sciences primarily to alert ecologists to the fact that new philosophical developments continuously take place and some may appear useful in forming a balanced and complete perspective of the issues involved in advancing ecology (Box 1.4).

We have found the new philosophy to be especially appropriate to ecology. Ecologists are impressed by the complexity of pattern and causality in natural systems (Roughgarden and Diamond 1986) and must deal with several "levels of organization" (Box 1.5), spatial heterogeneity (Kolasa and Pickett 1991), and an important role of history (Davis 1986, Foster and Aber 2004, Russell 1997). Indeed, many ecological and evolutionary systems are highly contingent on their past states and structures, as well as on current environmental conditions (Gould 1989, Kingsland 1985). In addition, ecological systems are rarely governed by one dominant cause, so the strategy for integration in ecology must deal effectively with interacting multiple causes (Hilborn and Stearns 1982). Multiple causality means experiments that are designed to discriminate between a single sufficient cause and ineffective agents may not identify a necessary cause and may in fact be misleading (Thompson 1989). Multiple causes are, together, necessary and sufficient for the phenomenon of interest.

The new philosophy of science has considerable power to advance ecological understanding by clarifying the goals of science, the structure of theory, and the relationships of various activities that scientists conduct. We will show how key philosophical insights apply specifically to understanding in ecology in Chapter 2. To highlight insights of modern philosophy that are particularly relevant to ecology, we present some characteristics of the new view of theory.

BOX 1.4 Constuctivism, Realism, and Postmodernism: Some Additional Philosophical Views

van Fraassen (1980) has proposed an interpretation of scientific process, labeled empirical constructivism, that has influenced some ecologists (e.g., Keddy 1992). According to empirical constructivism, holding a theory need not mean holding it to be true but can and should only mean holding it to be empirically adequate. A theory is empirically adequate when it does a good job of explaining or conforming to the observed, empirical phenomena. While empirical adequacy can be a criterion for judging a theory, empirical constructivism has been criticized for being too narrow in dealing with entities that are not observable and in ignoring the great interests scientists have in such conceptual entities. Furthermore, empirical constructivism seems to be at odds with how scientists actually work, particularly when it divides the world into observables and instruments — that is, entities that are not observable, such as living dinosaurs and electrons, but that are postulated for the sake of theoretical constructs. For example, "with his 2002 book van Fraassen has staked out much more radically antirealist ground, by denying that there is any justified distinction between lawlike and purely accidental generalizations" (cited after Don Ross, personal communication). Few scientists would find this to reflect the view of their discipline. The inadequacy of empirical constructivism will become apparent as we further explore the role of theory in achieving integration and understanding of ecological phenomena.

Another philosophical perspective of potential appeal to practicing scientists is known as scientific realism. Scientific realism represents a view of scientific theory which holds that (1) science aims to give us, in its theories, a true story about what the world is like and that (2) acceptance of a scientific theory involves the belief that it is true. This means that a scientific realist believes that the theory, observations, and nonobservable notions about nature are one, intertwined system of evolving knowledge, without any component being viewed as unnecessarily contrived or superfluous. The doctrine of scientific realism has emerged in response to more restrictive or problematic philosophical views. Scientific realism can be seen as attempting to remedy problems of (1) the empirist challenge we commented on earlier, (2) the neo-Kantian challenge (Hanson 1961, Kuhn 1970) that casts doubts on the long-term reliability of unobservable or theoretical components, and (3) the postmodern challenge, which views science largely as a social construction suggesting, therefore, that one should not hope for any sensible fit between the real world and such constructs (cf. Koertge 1998). In an attempt to defuse some of the more extreme criticisms of these three stances of each other and of science in general, scientific realism offers yet another philosophy of science closely allied with science itself. It appears to be well defended via arguments close to philosophical naturalism.

Theory is a large tool in science. Yet it is sometimes erroneously considered to be synonymous with one or another of its component parts. For example, quantitative models are often equated with the whole of theory (Shugart 1984). Likewise, we showed earlier how the inductive chain invites confusion of "theory" with "hypothesis" and "law." These things do not lead to one another in a linear developmental sequence. Rather, hypotheses and laws are components of theory. The statement view also drew an artificial boundary between theory and fact. The contemporary view recognizes that theories do, in reality, incorporate factual information in some

---

BOX 1.5 Traditional Levels of Organization for Ecology

Fine scale
Molecule
Subcellular structure
Cell
Tissue
Organ
Organism
Population
Community
Ecosystem
Biosphere
Coarse scale

---

of their components and that facts have a conceptual aspect. We will further develop these ideas in Chapter 3.

Once classical physics was displaced from its monopoly as the paradigm for philosophy and methodology of science, the real variety and dynamism of theory appeared. To undercut the position of physics in science is not our intention. However, a predator-competitor supercollider could bring about major progress in community ecology for a fraction of the cost of its physics analog. An especially important lesson from contemporary philosophy is that theory is not static. Rather, it changes and develops or is discarded, in part or as a whole, over time. Theory may start out as a series of vaguely linked notions and may expand to include a full array of components (Chapter 3). In addition, each of the components of theory may evolve from an imprecise and vaguely articulated form to a more exact and specific form (Chapter 4).

V. What an Integrated Ecology Might Look Like

A truly integrated ecology might have several important features, although it is currently impossible to draw a detailed map of such new territory. Our hope is that this unknown territory be treated as an opportunity rather than with the dread symbolized by the old map makers' phrase, "*Ibi dracones*" — "Here be dragons." Ecological integration might effectively bridge existing paradigms, link various levels of organization, and seek generalizations that cross currently disparate concerns.

A critical area of integration would be combining the two major paradigms in ecology already discussed (Fig. 1.4). A truly integrative ecology would deal effectively with the paradigm of matter and energy fluxes, here labeled the common parlance ecosystem paradigm, as well as with the paradigm of organism interaction, labeled the population paradigm, which includes community ecology. Important questions that lie at the intersection of these disparate paradigms include these: What is the role of taxonomic or physiological diversity among species in ecosystem function? How do controls on ecosystem nutrient flux limit species richness? What are the different ways species affect ecosystem functioning? Ecologists are actively pursuing these and many related questions (Jones and Lawton 1995).

Although it may not be possible to accomplish a complete integration of these two quite different paradigms, at least their points of intersection should be much better explored. Notably, both the organism and the ecosystem paradigm require the explicit use of history, including development and evolution, for at least some of the important questions. This is an important characteristic that differentiates them from the paradigm of classical physics, which does not require the history of its units to be known.

Recognizing that at least two major ecological paradigms exist suggests the existence of various nested hierarchies of ecological entities (Fig. 1.6), rather than the single classical hierarchy of levels of organization (Box 1.5). Each of the hierarchies in Figure 1.6 might be most appropriate to a particular paradigm or kind of question (MacMahon et al. 1978). Not all hierarchies may be appropriate to all research questions. For example, questions of resource partitioning among entities would relate to one hierarchy, whereas questions of phyletic relationships would relate to another, and questions about assimilated energy fluxes would relate to still another hierarchy (Allen and Hoekstra 1992, O'Neill et al. 1986, MacMahon et al. 1978, Salthe 1985). It is significant that all the traditional ecological hierarchies intersect at some points more strongly than at others. The individual organism is such a node, suggesting that this ecological entity may be a reasonable place to begin to explore commonality and integration in ecology (e.g., Huston et al. 1988). The fact that organisms possess a physiology, an evolutionary history and lineage, a taxonomic status, and influences on external resources suggests the importance of this node. Of course, populations, ecosystems, and communities share some of the same features. But other than via analogy, none of the other ecological entities commonly used share all those features.

Alternatively, the traditional view of ecological hierarchies can be replaced by a perspective in which various descriptions usually associated with the population and ecosystem paradigms become facets or aspects of a single hierarchy of ecological entities (Fig. 1.7; cf Allen and Hoekstra 1992). This hierarchy of entities offers additional advantages because it is scale independent and thus permits application of the same methods, concepts, and terminology across a broad range of more or less tangible ecological objects. For example, a distinct patch of forest can be studied as an ecosystem (physical dynamics of nutrient and energy fluxes), as a coevolving assemblage (evolution), as a web of interactions and relationships among species (community ecology), as a functional spatial pattern (landscape ecology), or in its behavior (succession). In the model of ecological approach and hierarchy (Fig 1.7), moving from one side, or facet, to another implies shifting research approach. Moving up and down the vertical axis representing scale implies moving shifting spatial extent and grain. For example, along the scaling axis concerned with space, the study unit can shift from a tree, to a forest stand, to forest patches, to a mosaic including forest and nonforest patches. Along the temporal dimension, seasonal contrasts, successional trajectories, paleoecological trends, or evolutionary time, can be discerned for a study system. Slicing the ecological pie vertically expresses a particular scale, and slicing it horizontally reveals different perspectives for studying ecological systems.

The facets are scale independent. They will always be present, but their expression and relative importance are likely to change with scale, often in predictable and regular ways. It is exciting to notice that these facets intersect at virtually any scale, a great departure from the hierarchy of MacMahon et al. (1978) which was an important advance when it was proposed. Unification in ecology may benefit from the full realization of this more realistic approach to hierarchy and from moving away from the one-dimensional textbook conceptualizations of ecological hierarchy (Box 1.5). There is no need to reduce a situation to the lowest possible hierarchical level to produce a successful explanation. We explore this advantage next.

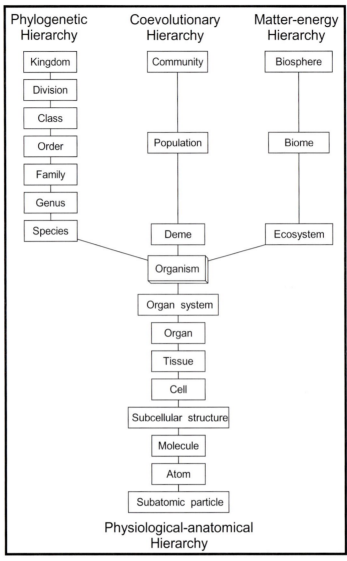

Figure 1.6   A refinement of traditional ecological hierarchies. Traditionally, ecological phenomena have been described as residing on a single nested hierarchy ranging from subatomic particles to the entire universe. MacMahon et al. (1978) refined the traditional view by dividing the hierarchy beyond the organism level, depending on the kind of question posed. One ecological hierarchy focuses on phylogenetic change, a second focuses on interaction of organisms within assemblages ("coevolutionary"), and a third focuses on the exchange of matter and energy. Within the organism, the hierarchy focuses on component structures, ranging down to the molecular and atomic. Adapted from MacMahon et al. (1978). Copyright 1978 by the American Institute of Biological Sciences. Used with permission.

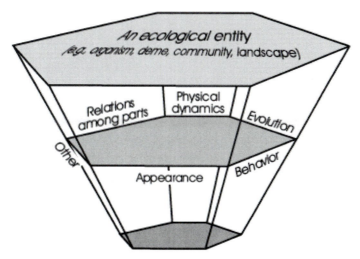

Figure 1.7 The various facets of ecological entities are the subjects of different research traditions. All these facets can, in relationship with the entity, scale up or down, and they can be applied to different hierarchical layers. The three hexagonal surfaces labeled "ecological entity" represent three nested hierarchical levels of ecological systems. We include a facet for "other" to indicate that the range of ecological concerns is broad, as we do not intend to exclude any legitimate ecological questions here.

Because ecological entities are organized hierarchically, integration in ecology must effectively deal with various levels of organization and the interactions across levels. Classically, a nested hierarchy of ecological entities has been used to explain the different kinds of ecology and to justify their being treated differently on the basis of "emergent properties" of their focal entities. Unfortunately, the traditional nested hierarchy, running from suborganismal units to the whole biosphere, can be considered to suggest that one or another ecological entity or approach is *the* correct one. The contemporary approach to hierarchy suggests a more unifying role for the idea of levels of organization in ecology.

Ecological questions can be asked at any level of organization, and the explanatory web can be cast upward and downward from that initial level. In developing ecological understanding in a hierarchical context, interactions or patterns will often be found on the level of immediate interest, whereas mechanisms for those interactions will be found at least one level below and explanations in terms of constraint and temporal and spatial context will often be found on the level above. For example, to understand the dynamics of vegetation in a plot of land, understanding processes at more inclusive and disaggregated levels is required. Population inter-action among plants and animals is the first level of disaggregation. But the dynamics of vegeta-tion that those populations compose can also be affected by neighboring patches of vegetation (Fig. 1.8). Processes at the higher level include dispersal of new propagules from the adjacent vegetation or alteration of the environment in the field as a result of the shading or other physical effects of adjacent vegetation patches. Therefore, to understand the focal process of vegetation dynamics, both disaggregation into component parts and placing the vegetation in a higher mechanistic level are required. The details of a hierarchical model for understanding any particular problem will be specific to that problem; the sketch just presented is only illustrative. We must leave discussion of the growing theory of organization and the hierarchical approaches

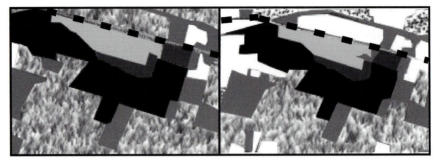

Figure 1.8 Maps of the Hutcheson Memorial Forest Center visually estimated from aerial photographs taken in 1962 (left panel) and 1995 (right panel). The old growth forest is shown as the solid black figure. Amwell Road is represented as the heavy dashed line to the north of the forest. Younger forests are shown in light grey. Post-agricultural old fields, studied by means of permanent vegetation plots since 1958 are shown in dark grey. Farm fields are indicated by the "grassy" hatch marks, and residential areas are shown in white. Young forest land and farm land were converted to residential use during the period mapped. General features of the successional study can be found in Pickett (1982) and an example of the process studies in this landscape can be found in Meiners et al. (2002).

to other sources (Allen et al. 1984; Allen and Hoekstra 1992; Kolasa and Pickett 1989). At this juncture, we offer a mini primer of hierarchy as a seed for further exploration (Box 1.6).

An integrated ecology would identify the deductive or causal laws and empirical generalizations of the discipline, which may also be viewed as laws (Colyvan and Ginzburg 2003). In integrating ecology, the goal will be to look beyond taxonomic limits, to link existing theoretical and empirical approaches to ecology, and to seek common processes and interactions in disparate ecological systems and circumstances. Too often, ecology is presented as a series of relatively isolated case studies. Of course, to generalize successfully from such studies, each case must have a specified and perhaps even a limited scope of application. In other words, an important task in constructing generalizations is to assess the limits, or domains, of generalizations and to identify the conditions under which they do and do not apply. Determining the limits to generalization will also suggest the conditions under which system contingency — that is, dependence on historical conditions of the system — is dominant. These problems are explored in several chapters, especially Chapters 6 and 8.

VI. Conclusions and Prospects

This chapter has used integration in ecology to identify the problems that motivate this book. Integration in such a pluralistic and diverse a discipline as ecology is a potentially powerful strategy for advancing the science. Ecology is replete with dichotomous debates, divergent scales, causal alternatives, and conceptual difficulties that can be solved by integration. The book explores the tools needed to conduct broad integrations in ecology. Enhancing integration in ecology can improve its effectiveness as both a basic and as an applied science.

The second major point of this chapter has been to provide some highlights of the contemporary philosophy of science that enhances our ability to turn the tools and methods of science toward integration in ecology. Few ecologists realize that the most commonly cited and discussed

---

BOX 1.6 What Hierarchy Is and Is Not

Most ecologists agree that ecological systems are hierarchical, but many have difficulty applying the notion to any but the simplest of situations. As the concept of hierarchy applies to the subject matter as well as to the structure of theories, we supply a few clarifications. This box might be used, for instance, to practice formulation of conceptual hierarchical models of known ecological systems, whether spatial, organizational, or process oriented, or to analyze the structure of existing models or theories.

Definitions

1. The organization of people at different ranks in an administrative body (www .hyperdictionary.com)
2. A series of ordered groupings of people or things within a system; "put honesty first in her hierarchy of values" (www.hyperdictionary.com)
3. Hierarchy is the condition being composed of (nested) subunits (Kolasa and Pickett 1989)
4. The term "hierarchy" is often applied to any representation of the hierarchical structure or ordering of parts, concepts, or levels (Salthe 1991)
5. Hierarchy is a partial ordering of sets (Bossort et al. 1978)

Interpretations

All of these definitions require that hierarchy be identified in the context of a system. First, one needs to specify what a system is (administrative body, groupings of things within a system, an array of sets, or a thing being composed of something). Only then it is possible to reflect on hierarchy.

---

philosophical insights, those of Popper and the logical positivists, are problematical and have been superseded in the contemporary philosophy of science. The classical philosophy emphasizes theories as statements, addresses disciplines that are both theoretically mature and characterized by simple causality, and promotes a single methodology for all sciences. The approach to confirmation advanced by logical positivism and Popper's 1934 (translated into English as Popper 1968) proposal that falsifiability solved the problem both unduly restrict the methods available to scientists in diverse disciplines. The issue of strong inference is hobbled by the same problems as logical positivism. All these methodological dicta fall with the failed view that scientific theories are deductively related series of statements. The contemporary philosophy of science provides a new view of the nature of theory, suggests the joint conceptual and empirical content of theory, and shows the multiple ways in which theory and observed phenomena can interact. The remainder of the book uses these key insights from contemporary philosophy of science and applies them to integration in ecology.

How can ecologists enhance integration? We believe that theory — with its demand for conceptual clarity and statement of assumptions, its provision of models of how ecological systems are put together and how they work, its potential for unification and generalization, its rich empirical content, and its frameworks for tying all these features together — is likely to be the most effective tool. However, theory is only a part of the scientific dialogue with nature that generates understanding. To enhance integration in ecology, we must first comprehend the nature of understanding in ecology and its relationship with other tools and tactics used by scientists (Chapter 2). Part II examines the nature of theory in more detail. It presents and characterizes the components of theory (Chapter 3), shows the richness of kinds of theory and how they develop (Chapter 4), and shows how theories differ according to their objectives (Chapter 5). Part III explores how changes in understanding are stimulated (Chapter 6). Knowledge of how understanding changes permits us to outline how integration can develop (Chapter 7). Theory is used in a context (Part IV) composed of both scientific worldviews (Chapter 8) and social factors (Chapter 9). These two contexts constrain and shape integration. The challenge to ecology is great, but its foundation is broad and firm.

2

Understanding in Ecology

I. Overview

We have identified increasing integration in ecology as an important and urgent need. To achieve this goal, we need ways to both evaluate and facilitate integration. In this chapter, we examine understanding as a necessary state for achieving integration. Indeed, understanding is the overarching goal of any science. This chapter defines understanding in terms that are useful to practicing scientists and relates that intellectual state to the process by which it is achieved: that of general explanation. We also show how other, more specific tools and activities contribute to understanding. The specific tools for constructing understanding are causal explanation, generalization, and testing. Testing can rely on both confirmation and falsification. In addition, understanding requires delimiting a domain of discourse in which the other tools are then applied. Understanding has two components that are briefly introduced in this chapter: observable phenomena and conceptual constructs. An iterative dialogue between these two components is the principal method by which understanding develops and changes. The chapter closes with an examination of the diverse realms in which ecologists work. These different realms engage a variety of tools and components of understanding. All of these tools and activities that lead to understanding touch in some way on theory, which will become the principal theme throughout the remainder of the book.

II. The Nature of Scientific Understanding

Scientists are fond of discussing what they do and what constitutes "good science." Fortunately, historians and philosophers of science have covered the same territory well and deeply for a long time. We have found key parts of this scholarly tradition to be helpful in setting the stage for considering the role of integration and theory in ecology. The consensus among philosophers is that the most general goal of science is to generate understanding (Burke 1985, Crombie 1953, Ruse 1988, Salmon 1984). Understanding here has a specific meaning that we must expose. In a scientific context, the term "understanding" implies that questions about a phenomenon can be answered by referring to certain patterns in nature, relationships among entities and processes, and causes of the patterns and their differences. In other words, understanding puts new knowledge in the context of existing knowledge. For instance, to say that we have an ecological understanding of insect communities on plants (e.g., Strong et al. 1984) suggests a domain consisting of insect abundance, co-occurrence, and diversity on plants. This statement suggests knowledge of the causes, such as resource limitation, plant architecture, resource competition among insects, and the agents and effects of abiotic and biotic mortality. It suggests knowledge of larger scale patterns, such as those of species-area relationships, regional species pools, and plant distribution and abundance. It also indicates that we have some expectation of the circumstances under which insect species richness, for example, will be high or low. Some of this information might be considered useful for pest management or for conservation.

In a similar vein, to posit an understanding of rocky intertidal communities (Lubchenco and Menge 1978, Sousa 1985) suggests knowledge of their spatial patterns along the elevation gradient spanning low to high tides, patterns of community change through time at specific sites, and differences in community richness and structure among geographic regions. Furthermore, the relative importance of three kinds of phenomena — biotic interactions, physical stresses associated with tidal fluctuations, and the impact of wave-generated disturbance — can be used to explain the spatial and temporal patterns of these communities. These three factors can also be used to generalize existing findings to new sites, regardless of whether those sites are similar or different (Sousa 1985). The understanding of rocky intertidal communities can also involve the application of ideas developed initially under quite different circumstances, such as the concepts and expectations of succession theory (Miles 1979). Likewise, the explanation of rocky intertidal community patterns and processes suggests comparison with the patterns and processes influencing communities in soft-bottom intertidal communities (Peterson, 1991).

To move to another area of ecology, an understanding of nitrogen flux in an ecosystem (Fig. 2.1) would involve knowledge of the spatial and temporal patterns of inputs and outputs, nitrogen pools, nitrogen transformations and transfers between pools, and the circumstances that cause net release or immobilization of nitrogen in various compartments within the ecosystem

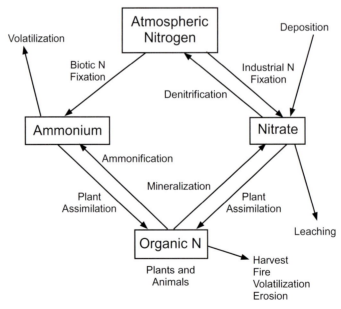

Figure 2.1   A simple nitrogen cycle for an ecosystem. Pools are shown for the atmosphere, nitrate, ammonium, and organic forms of nitrogen. Transformations between the pools driven by biological, physical, and industrial processes are shown and named. Because the model is conceptualized for a single ecosystem, inputs and outputs from the system are also shown.

or in the entire ecosystem (see, for example, Schlesinger 1991). Furthermore, for a complete understanding of nitrogen dynamics, we must know the functional groups of organisms involved in consumption or release of nitrogen, how those groups compete with one another for nitrogen, and how they depend on the transformations of nitrogen by other groups of organisms in the ecosystem (e.g., Risser 1988). Comparisons of such patterns and their controls in contrasting and similar ecosystems can lead to a general understanding of the patterns and causes of nitrogen fluxes (Vitousek and Matson 1991). The understanding of N dynamics also requires knowledge about human-induced inputs of nitrogen and anthropogenic influences on its processing. Nitrogen flux in ecosystems is frequently affected by human activities. For example, physical disturbance of soil often stimulates nitrification. Logging converts leafy trees, which are a relatively rich reservoir of nitrogen, to litter and debris without the opportunity to resorb N from leaves at the end of their life cycles. Alteration of groundwater tables, organic matter, and aeration of riparian zones can reduce the ability of those streamside fringes to convert nitrate pollution to harmless nitrogen gas. Therefore, the implications for management, such as logging methods, buffer zones around suburban development, or farms to protect riparian zones or wetlands, and the design of conservation areas, may all emerge from an understanding of ecosystem-level nitrogen fluxes (e.g., Ehrenfeld and Schneider 1991).

Understanding may take place at different levels of generality. For example, an ecosystem ecologist may need to know the contribution of individual species or functional groups to nitrogen flux before reaching a satisfying level of understanding. In contrast, an undergraduate student in an ecology course may only need to understand that biota in certain compartments move or transform nitrogen at varying rates depending on their identity and activities, without knowing any of the underlying details.

A. Definition and Components of Understanding

The examples outlined share certain characteristics that collectively suggest a scientifically useful definition of understanding (Campbell 1974b, Fagerström 1987, Hoffmann 1988, Levin 1981, Mahner and Bunge 1997, Salmon 1984, Stegmüller 1976, Suppe 1977a). *Understanding is an objectively determined, empirical match between some set of confirmable, observable phenomena in the natural world and a conceptual construct.* In other words, understanding is a state that refers to the degree of match between reality and theory, a match between what scientists observe and what they think (Fig. 2.2). De Regt (2004) augmented this interpretation by adding the pragmatic value of understanding — the ability to account for a natural phenomenon by using the theoretical knowledge in the preceding definition. To better appreciate the nature of scientific understanding, we will examine its components and present some examples. Note that scientific understanding is, in some ways, like other modes of understanding in that a picture of the world is generated to represent the world (Box 2.1). However, some of the modes of understanding outside the realm of science (e.g., poetry or music) involve rather personal and idiosyncratic pictures of the world, whereas scientific understanding is generated as a public process. Furthermore, these other modes of understanding are not intended to be evaluated or improved in the same way that scientific understanding is.

Few discussions of the general nature of theory in philosophical are cast in terms that are relevant to ecologists (but see Thompson 1989). Of course, a plethora of successful and important specific ecological theories, and excellent treatments of these, exist in ecology (e.g., Anker 2002, Ford 2000, Hubbell 2001, Roughgarden et al. 1989, Turchin 2003, Ulanowicz 2004, Yodzis 1989). However, we believe that one of the surest ways to enhance our understanding in ecology and, consequently, to promote integration of the discipline, is to make the inclusive nature and wide utility of theory in its most general sense better known and comprehended by ecologists.

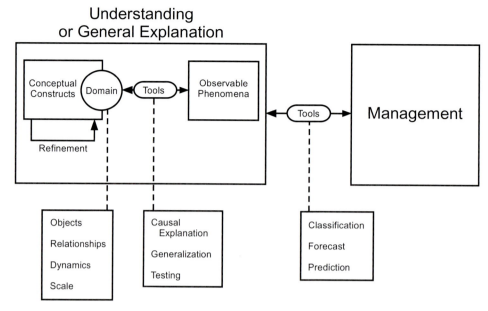

Figure 2.2   Understanding and the tools for establishing or using it. The state of understanding and the process that
generates it, general explanation, are represented by the largest box, containing their component conceptual constructs and observable phenomena. These two components are interrelated by a series of tools
that are exercised within a specified domain. The fruits of understanding can be applied, through still other
tools, to management or policy concerns raised by society.

Therefore, the components of theory as well as their use, refinement, and replacement are the
central themes of this book.

Given the technical definition of understanding given earlier, we can now expand on its components. Understanding can be envisioned as a structure supported by two pillars — one conceptual and the other empirical. The first pillar or component of understanding is the conceptual
constructs against which reality is to be matched. Some concepts refer to simple and straightforward features of nature. A relatively simple concept is "tree." The concept "tree" is distinct from
that of "maple" or "oak," but it identifies features common to both specific kinds of tree. Other
concepts can be highly abstract and constructed or derived from simpler concepts. For example,
a model of tree growth is a more abstract and derived conceptual construct than the concept of
"tree." Conceptual constructs of varying degrees of complexity are collectively assembled to
create "theory" in the most general sense. Conceptual constructs can and usually do include
empirical facts. However, the selection and inclusion of these facts or data are guided by their
usefulness to the explanatory task at hand. Conceptual constructs go beyond facts because their
function is to generate understanding of a new class of observations not previously understood.
We will examine this blending of empirical facts and concepts at greater length in Chapter 3.

The second pillar of understanding is the observable phenomena that ecologists wish to explain
(Fig. 2.2). This empirical pillar of the edifice of understanding is well illustrated by the contents
of various general and specific textbooks of ecology and by advanced reviews. Because of the
wide availability of this information, we will repeat only as much of it throughout the book as
is needed to exemplify our general themes.

The third component of understanding is the beam that rests on and connects the two pillars:
conceptual constructs and observable phenomena. This component comprises the tools by which

---

BOX 2.1 Modes of Understanding and Their Characteristics

These modes of understanding represent three important and contrasting ways in which humans make sense of the diversity of experience.

Via Science

Outcome: Conclusion
Features:
Replicable or confirmed observation
Community debate and test
  Empirical evidence
  Linkages of observation and explanation
Verification through replication
Tentative conclusions in a revisable system
Creativity

Via Faith

Outcome: Belief
Features:
Singular observation or experience
Community confirmation
Verification through affirmation
Fixed system

Via Art

Outcome: Expression
Features:
Singular experience
Community reception
Personal interpretation
Creativity

---

scientists relate the observable phenomena of nature to their conceptual constructs (Fig. 2.2). There is little or no scientific meaning to isolated conceptual constructs or unconnected observations of the material world. Science grows from the links made between these two components of understanding. It is the tools for understanding that facilitate a dialogue between concepts and phenomena. These tools are causal explanation, generalization, and testing (Fig. 2.2), which will be examined in greater detail later in this chapter.

The three general components of understanding introduced earlier describe what the dialogue is between — the concepts and observables — and how the dialogue is conducted — using the tools. However, like a conversation, this scientific dialogue has to be about something — that is, it must have a clear focus. Therefore, the dialogue must always occur within a specified domain.

The domain is the set of objects, relationships, and dynamics at specified spatial and temporal scales that are the subject of scientific inquiry. For example, the domain of population ecology includes growth, reproduction, survival, immigration and emigration, and other interactions among individuals at spatial scales that can explain changes in the density of a collection of individuals of the same species. This domain is informed by the monospecific ecological entity defined as a population. However, if we wish to explain changes in density of individuals of other species as well, then the domain is no longer that of population ecology but of community ecology. Similarly, if we do not seek to explain changes in density or other attributes of a *collection* of individuals of the same species but instead seek to explain patterns of individual growth or survival, then the domain becomes autecology, not population ecology. In a similar vein, ecological studies that confine the temporal dimensions to time scales that exclude evolutionary processes are in the domain of ecology, sensu strictu. When time scales encompass such evolutionary processes, the domain becomes that of evolutionary ecology. While these delineations do not always appear sharp, it does not change the fact that domains exist and play a crucial role in advancing the dialogue between the components of understanding. Without explicitly specifying the domain, scientists can reach no agreement about when understanding has been achieved. Consequently, domain will receive considerable emphasis in this chapter and in Chapter 3.

One of the key phrases we employed in our definition of understanding was "objectively determined, empirical match." We must pause to explain objectivity here, because it has been a persistent thorn in the side of philosophy. We cannot resolve the philosophical debate, but we will report recent philosophical insights that we believe are scientifically appropriate and useful to this issue within ecology. An "objectively determined, empirical match between observable phenomena and conceptual constructs" means that the degree of match is openly constructed and evaluated by a community of scientists (Hull 1988, Longino 1990). The dialogue involves free access using a shared public language that may be verbal, quantitative, or qualitative. Measurements of a phenomenon or class of phenomena are repeatable or confirmable by other observers. Unique events are confirmable if their subsequent effects are subject to scientific analysis or if they are studied by several independent researchers. An objective stance recognizes that the match between reality and conceptual constructs is tentative and subject to revision or replacement in the scientific forum as a result of new data, explanations, concepts, and tests. These assessments are themselves open to analysis and criticism by the community. Objectivity results from the cancellation of individual bias by the active participation of a diverse community of scientists in an open-ended interrogation of nature (Longino 1990). The cancellation of bias is important because scientists may both intentionally and unintentionally introduce bias. Biases may arise from many sources, including scholastic tradition, funding, intended audience, methodology, and unrecognized assumptions (Taylor 2005). Therefore, the identification of bias is important, and the operation of an open scientific community that effectively cancels both recognized and unrecognized biases is crucial.

The old philosophy of science focused on the individual scientist's behavior as the way to ensure objectivity. This view has now been de-emphasized as the role of the entire community as a necessary feature of the scientific process has become clearer (Hull 1988). The manifold implications of this new philosophical insight will become apparent as the structure of theory, its development, and its replacement are discussed in subsequent chapters.

B. Relationships between Understanding and Explanation

We have taken understanding as the primary and most inclusive goal of science. This perspective may seem to contradict the statement by some philosophers and historians of science that explanation is the most general goal of science (Miller 1987, Weissman 1989). The contradiction can be resolved because "explanation" is the *act* of relating some phenomenon to a conceptual construct (Salmon 1984), while understanding is the *result*. The act of explanation can also be described as putting the unknown or novel into the context of the known and expected and, hence, into the context of some specific or general model (Lloyd 1983, Pielou 1981). So explanation is an act or *process*, whereas understanding is the resultant relationship or *state* (see also Ford 2000, p. 272).

Explanation is often used in somewhat different ways in science and philosophy. In the classical philosophy of science, explanation has been supported by a series of general laws (Carnap 1966, Nagel 1961). More recently, philosophers have taken explanation to mean discovering the relationship of a pattern or phenomenon to its causes (Mayr 1961, Rensch 1974, Salmon 1984, Scriven 1959, Suppe 1977a). A specialized way to relate patterns and causes is to break a phenomenon down into other phenomena at lower hierarchical levels of organization (Pattee 1973). For example, photosynthesis is explained by biochemical and physiological processes, leaf anatomy, fluxes of light, gasses, and water. Nitrogen fluxes in an ecosystem are explained by changes in the rate processes such as fixation, mineralization, and immobilization. Philosophers sometimes refer to this process as *microexplanation* (Mahner and Bunge 1997). At the same time, explanation may often require putting some specific phenomenon into the context of broader scale phenomena or events that enable or constrain the phenomenon of interest (O'Neill et al. 1986). For example, rates of photosynthesis may be constrained by nitrogen availability at a site, and local nitrogen fluxes may be enabled by nitrogen inputs from the larger landscape. This is philosophical *macroexplanation* (Mahner and Bunge 1997). Explanation can also come in other flavors (Mahner and Bunge 1997). *Statistical* explanation involves employment of a probabilistic law statement to explain a phenomenon, while *narrative* explanation (Gaddis 2002) involves implicit assumptions, logic, and conjectures such as those used in interpretation of evolutionary patterns.

These more specific uses of the term "explanation" — exposing underlying mechanism, placing phenomenon into broader scale context, and using probability or narratives — are clearly subsets of the general process of relating conceptual constructs to observable phenomena (Fig. 2.2). We will, therefore, use the term "general explanation" for the act of constructing broad understanding of a large domain, and we will refer to putting specific phenomena or events into the context of laws, causal frameworks, or hierarchies as "causal explanation." These two uses of explanation must be kept distinct.

III. Toward Understanding

A. The Component Processes of General Explanation

In this section, we will flesh out the processes that contribute to understanding and will relate them to one another. We have already noted that two key components generating understanding are (1) the domain, or the bounded universe in which the dialogue between conceptual constructs and observables is conducted, and (2) the set of tools that relates the conceptual constructs to the observed phenomena. We will discuss these two components of understanding and their role in the process of general explanation.

B. Domain

While we have briefly illustrated the nature of domain and its critical importance, it can be further illustrated using several well-known ecological theories. In island biogeography, the domain includes these entities or processes: species and records of presence and absence of species; an ecological, rather than geological or evolutionary, time span; truly isolated islands; aggregated processes of extinction; aggregated processes of immigration; physical parameters of distance and size of islands; and a source area for species (MacArthur and Wilson 1967). Choices in specifying the domain may include the option to use island analogs on land, which can sometimes be problematical. The domain of island biogeography clearly does not include organismal physiology, population dynamics, processes of species replacement, competition, predation, or nutrient cycling. These phenomena have been used commonly in other theoretical realms concerned with the structure and persistence of biotic communities.

For community succession, the domain includes an assemblage of species; characteristics of species such as life form, reproductive strategy, and competitive ability; initially open sites for species interaction and the resource and stress levels associated with those sites; and a suite of general processes, including site opening by disturbance, species invasion or persistence resulting from disturbance, and species interactions (Pickett et al. 1987, Pickett and Cadenasso 2005). Choices and problems with the domain of succession that may lead to controversy or contradiction include whether to focus on ecosystem parameters as drivers and the spatial and temporal scales of focus. Examples of narrow successional domains sometimes inappropriately considered to be distinct are natural forest regeneration, production forestry, postagricultural old-field succession, and primary succession (Luken 1990). The domain of community succession clearly does not include phenomena of biogeographic extent, questions pertaining to species evolution, or energy budgets. It can, however, be linked to these other domains as a target for integration.

The coevolutionary theory of plant-herbivore interaction includes the following in its domain: higher plants and their secondary metabolism, the identity of the herbivores and their degree of host specialization and digestive physiologies, natural selection for defensive attributes of plants by insects, selective herbivore adaptation, and evolutionary time scales (Ehrlich and Raven 1964, Thompson 1982).

As we mentioned earlier, a domain may somewhat overlap with another domain. For example, the coevolutionary theory of plant-herbivore interaction overlaps with domains addressing plant adaptations to abiotic sources of plant stress and damage and microbial pathogens (Coleman and Jones 1991, Coleman et al. 1992, Jones and Coleman 1991). Hence, a strict delineation of a domain may not be possible and not even desirable in some cases. However, the notion of domain assists in organizing a discourse on a particular set of phenomena and ignoring it will almost always lead to unproductive, confused debates.

Several themes emerge from even this brief glance at the domains of ecological theories. First, ecological domains have most often focused on entities rather than processes, since entities are easier to detect or measure in the field than are processes. The entities most ecologists commonly focus on are specific taxa. However, the definition of ecology indicates that a principal focus ought to be on interactions rather than on entities (Chapter 1). Second, there is a diversity of domains within ecology, even within one topic area such as succession. For example, the domain of Clements's (1916) original synthesis of plant succession was confined to changes in life form. His insistence that succession only progressed in one direction became problematical when discussion was extended to species composition.

It is possible to choose to focus on rather discrete corners of the universe of discourse. Therefore, several caveats can be offered about domain: (1) domain must be specified as explicitly as possible, otherwise important assumptions about the nature and function of a phenomenon may remain hidden; (2) domain can expand as new processes or structures are discovered or can contract as a theory is found to fail in some specific topic area in a larger universe. As examples of expanding domains, we can cite the expansion of island biogeography to consider host plants as islands (Janzen 1973) and the expansion of community ecology to include multiple interactive communities as a metacommunity (Leibold et al. 2004). Likewise, recognizing the fundamental similarities of interactions between plants and herbivores and plants and pathogens required a domain expansion (Barbosa et al. 1991, Colman et al. 1992). Domain expansion is currently underway in the study of ecological boundaries, so that soils, surface, and aquatic systems are all now included, and transfer processes are examined over an immense range of scales (Cadenasso et al. 2003). As an example of domain contraction, coevolutionary theory sensu strictu refers to stepwise reciprocal adaptation, as opposed to a broader and expanded domain of diffuse coevolution encompassing evolution of multiple interacting players. (3) Spatial and temporal scale and hierarchical level of organization are critical aspects of the scale of a domain. Failing to specify them may result in misapplication of theories or models. Because domain is so fundamentally important for effective dialogue between conceptual constructs and observable phenomena, we will consider it one of the components of theory (Chapter 3).

C. Tools for Understanding

We now discuss in more detail the set of tools for relating conceptual constructs with observable phenomena. These are the tools for building understanding within a domain. We will discuss the tools in order, from those that emphasize pattern generation, to those that emphasize formulation of causal explanations, to those that test the relationship between pattern and cause, and finally to those that test the validity of causal explanations.

1. Generalization

Generalization often involves pattern generation. This step is critical in ecological research because without pattern, we cannot determine the significance of a prospective explanation. The diversity of biotic and abiotic components and the variety of ways they can interact in nature mean that many outcomes are possible, yet only some actually become expressed as pattern. Thus, pattern constrains us to address what we see, not what we think could be. For example, the simplest of population dynamic models can generate an almost infinite variety of dynamic behaviors. While natural populations do exhibit many dynamic behaviors, they certainly do not show all possible behaviors. In some ways, generalization from pattern is a central justification for empirically based field research and a reason why some ecologists are dubious about what can be learned from "artificial" studies such as abstract models, controlled environment microcosms, or constructed species assemblages. Artificial studies are clearly of value in exploring and revealing possibilities, but pattern in nature tells us what is actually realized.

We begin our exploration of generalization by discussing relatively concrete sorts of generalization, and we will progress to those that are more abstract. The most concrete mode of generalization *condenses* many observations thought to be of a similar sort into a briefer summary statement, be it an equation, a graph, a sentence, or a numerical value. Mean monthly temperature is an example of a condensing generalization. Such condensing generalizations may highlight some aspect of nature that is in need of causal explanation and may therefore lead to the development of theory or one of its components. For instance, the observations of morphology, odor, shape, and color of flowers have been generalized into character syndromes appropriate for different pollen vectors (Fig. 2.3). These kinds of generalization describe patterns or processes in the material world. Such descriptive generalizations emphasize that there are many forms generalization can take, from narrative, through diagramatic, to quantitative, for example. Important ecological phenomena that emerged from such descriptive generalizations include community succession, food web efficiencies, and the associations of invasive species.

In some instances, if a generalization is made on the basis of relatively limited observations, it may be considered a hypothesis. Such generalizations can be treated as predictions — that is,

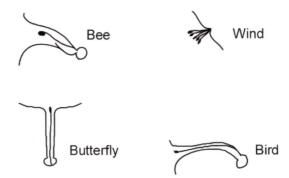

Figure 2.3   Pollination syndromes. Flowers have contrasting pollination syndromes, or kinds and relationships of physical and resource characteristics, that differentially attract and exclude pollinators with different behavioral selectivity, resource requirements, modes of locomotion, and so on.

as statements of expectation about additional cases separated in either space or time from the ones on which the original generalization is based. For example, the early discovery of a syndrome of anatomical and morphological characters associated with the then novel C4 photosynthetic pathway became a hypothesis about the relationship of the metabolism to the anatomy (Black 1971). The anatomy was expected to coincide with the metabolism; this was subsequently shown to be a valid generalization.

Generalizations about phenomena on a particular level of organization can stimulate mechanistic research on a lower level of organization. The admittedly speculative generalization based on limited observations that early successional communities display periods of alternating low and high heterogeneity suggested hypotheses about alternating periods of competitive exclusion and fine-scale community disruption (Armesto et al. 1991). Similarly, the observation that insect outbreaks sometimes correlated with drought periods led to mechanistic studies on effects of drought-induced changes in foliar nitrogen and defenses as potential causes of changes in pest performance (Waring and Cobb 1992).

Generalization is clearly a multifaceted process in the dialogue between observed phenomena and conceptual constructs. Many kinds of generalization and ways to employ generalization exist. However, certain features of generalization are particularly valuable. Quantitative generalizations are especially valuable because they are clear about exactly what is to be compared, and they improve the likelihood that the comparison will be in equivalent units and form (Keddy 1989). They can also be subject to statistical evaluation; the goodness of fit and the proportion of variation attributable to the independent variable can be assessed if the generalization can be cast as a regression (Peters 1980).

In many areas of ecology, quantitative generalizations should be extremely valuable. Unfortunately, many existing relationships are erroneously considered to be historically beyond the phase of constructing quantitative generalizations. For instance, many of the generalizations about species diversity and latitude or other coarse-scale environmental features are quite old, having been generated in an era where quantification was not considered paramount or was not technically feasible. As a result, many of these general relationships are qualitative or even anecdotal (Schimper 1903). The value of revisiting such relationships from a more quantitative perspective that then suggests new hypotheses and mechanistic studies is illustrated by the relatively recent statistical discovery of the nature of coarse-scale species richness relationships of trees in North America. Currie and Paquin (1987) found that 76% of the variation in species richness was attributable to annual evapotranspiration, suggesting a limitation of species richness by total energy flux. Data from Great Britain and Ireland fit the same relationship as the North American data. Numerous additional questions are suggested by the pattern presented by Currie and Paquin (1987): On what scale do other phenomena begin to correlate best with species richness? Do similar patterns of richness and actual evapotranspiration hold on other continents and biomes? Is the pattern true irrespective of the history of the region (e.g., South African fynbos communities) or fertility of soils? Similarly, the quantification of the pattern between insect species richness and geographic range for plants of different growth forms (Lawton and Schroder 1977) led to important insights into the role of plant architecture (Lawton 1983). Such coarse-scale ecological patterns — labeled macroecology by Brown and Maurer (1989) — deserve serious quantitative attention.

While quantitative, statistically valid generalization has real value, it is important not to over-emphasize it to the exclusion of nonquantitative condensation. Extremism and narrowness of method inhibit scientific progress. In this regard, Peters (1991) suggested that "predictive ecology," an avowedly empirical method of generalization based on regression and purportedly untainted by theory, is superior to other ecological approaches. We agree that pattern generation is a necessary ingredient in advancing ecology and stimulating integration. When this strategy is used, there are real advantages to careful choice of variables and sound quantification. However, slavish and exclusive adherence to that approach alone would be as damaging as exclusively applying any other single narrow approach in ecology. The usefulness of statistical generalizations as hypotheses about generality of pattern beyond the bounds of the data set used to generate them, and as sources of hypotheses for mechanistic work, is often considerable. However, the best use of statistically generated patterns will be in association with causal explanation and other conceptual constructs, as described subsequently. Pretending that an ecologically interesting regression is devoid of a connection to ecological theory is a sham. We believe it is better to admit the relationship so that the ecologically relevant assumptions and connections with other germane processes and interactions can be evaluated and brought to bear. For instance, note that even simple generalization by condensation of observations makes assumptions about the similarity of those observations, based on other knowledge about a topic. Such knowledge is codified in a relevant, even if rudimentary, theory. For instance, the successful "predictive" (sensu Peters 1986) relationship between lake productivity and phosphorus is based on a sound theory of photosynthesis and the factors that limit it in aquatic systems. Combining that theory with the recognition that phosphorus loadings differ among lakes underwrites the regression. Likewise, the relationship between tree richness and actual evapotranspiration (Currie and Paquin 1987) would have little ecological weight were it not for the theories of physiological ecology that underwrite the causal connections between tree performance, water use, and solar energy (e.g., Fitter and Hay 1987).

Now that we have presented several of the basic kinds of generalization, we can analyze important aspects of generalization as a process. These insights introduce the most abstract form of generalization. All generalizations involve simplification. The material world is too complex and various to comprehend or work with in one bite. Simplification is a necessary tool of science. One important aspect of simplification is *abstraction*, the identification of the essence of the phenomenon or interaction of interest (Fagerström 1987). Nonessential features of the system or interaction are ignored in constructing the generalization. For example, to generalize about the control of species richness by area, the identity of species from region to region is not considered important. This choice means that individual species characteristics are ignored; the important feature is considered to be the number of species, not their identities. This simplification allows scientific focus on only the form of the line, and the parameters of the equation for the species-area curve, to compare and apply the relationship (Fig. 2.4). Of course, for other

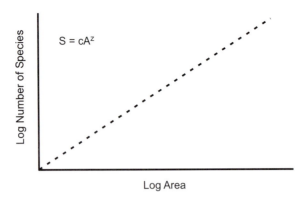

Figure 2.4   The species-area curve. The log species richness versus log island area plot is shown with the general equation that describes it.

purposes, species identity is critically important, but not for the particular generalization at hand.

A second aspect of the process of generalization is *idealization* (Fig. 2.5). This is another kind of simplification. Often, to generalize successfully, it is necessary to simplify the system by discounting influences that might in fact act on it (Levins 1966). Succession might not have been discovered as a general phenomenon in ecology if all the site- and history-specific accidents that can determine change and state of vegetation had not been ignored in doing the initial studies and making the largest comparisons. For example, upland succession in the North Carolina Piedmont contributed substantially to the development of the discipline. However, the patterns of community change in nearby lowlands were unclear from a successional perspective (Oosting 1942). In essence, they were left out of the idealizations to permit the field of study to consolidate. Likewise, identifying consistent features of trophic interactions in ecosystems would not survive the mass of detail on the diets of individual organisms. To be sure, important refinements are emerging in the study of food webs, but the field still depends on idealizing feeding connections to a great extent, using, for example, "trophospecies" or aggregates of species of similar trophic roles to construct and analyze properties of food webs (Martinez 1991, Williams and Martinez 2000, Yodzis 1993). Similarly, simplification via idealization is apparent in Grime's C-S-R triangle theory of plant community ecology, which forces all observed cases into three gradients of stress, disturbance, and competition (Wilson and Lee, 2000). An appropriate illustration of idealization from outside ecology appears in the area of classical mechanics. The laws of motion assume ideal behavior of bodies to see the common features of motion and interaction rather than the confusing details of friction or wind resistance. Complicating factors such as friction and wind resistance are taken into account in certain applications, of course, such as designing airplanes.

The most abstract feature of generalization is *unification*. Although condensing generalizations extract information from a set of observations that are considered similar, unification addresses domains that are initially thought to contain disparate phenomena. It is the generalization across difference that characterizes unification. The recognition that the principles of sessile animal defense in marine systems are the same as those of terrestrial plant chemical defense (Hay 1991) is a good example of ecological unification. Treating forest stands, mountain habitats, lakes, caves, dung piles, city parks, or islands as sharing the same general feature of isolation is another successful case of unification (e.g., Lomolino and Smith 2003). The recognition of the physical

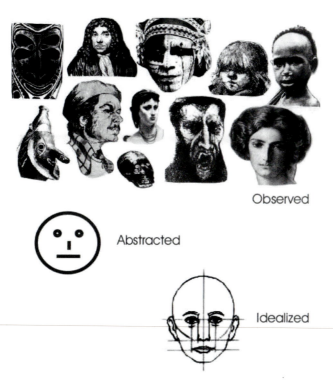

Observed

Abstracted

Idealized

Figure 2.5　Observation, abstraction, and idealization, illustrated using representations of human faces. In the first step, a researcher assembles the collection of observable phenomena or representations according to a particular research protocol (simple curiosity, in this informal case). Then, the researcher abstracts the collection to retain universal and essential features. In the last step, the research eliminates various constraints to render an idealized image (example of a flawless face as idealized by Leonardo da Vinci). In most cases, the idealized representation will be simpler than the abstracted one. Influences of a current social mindset (an equivalent of "paradigm" in science) on the idealization are obvious and may be provocative to contrasting mindsets or aesthetics. In other words, different idealizations of the representations presented in the figure are possible based on different aesthetics. In this light, da Vinci's idealization is clearly an expression of an ethnic and cultural time and place.

ecosystem engineering process (Jones et al. 1994, 1997) as organismally induced, structurally mediated changes in the abiotic environment is a unification. The generalization holds despite the myriad forms (e.g., beaver dams, tree shade, earthworm burrows, and mollusk shells) and its diverse consequence in organismal, population, community, ecosystem, and landscape domains. These phenomena were all originally considered to be distinct before the concept of ecological engineering brought them together. Outside ecology, the unification of electricity and magnetism is an example, as is the unification of space and time in general relativity.

2. Causal Explanation

Causal explanation is another of the tools for constructing understanding. As we have already said, it differs from *general* explanation, which refers to all the threads in the dialogue relating conceptual constructs and observable phenomena. *Causal explanation* is one of those threads. It refers to the determination of the processes, mechanisms, interactions, or conditions that result in a pattern or phenomenon (Box 2.2). In biology and ecology, causal explanations may reside

---

BOX 2.2 Targets and Aspects of Causal Explanation

Phenomenon. Any observable events, entities, or relationships of interest to ecologists.
Patterns. Repeated events, recurring entities, replicated relationships, or smooth or erratic trajectories observed in time or space.
Process. A subset of phenomena in which events follow one another in time or space. These may or may not be causally connected.
Mechanism. A subset of cause, referring to a direct interaction that results in a phenomenon.

---

in a wide variety of realms, ranging from biogeochemical and biochemical, through physiological and morphological, to the behavioral, community, ecosystem, landscape, and evolutionary (see Fig. 1.2). This causal breadth is one reason that ecology covers such an extensive subject matter and why ecologists often get into futile debates about what constitutes true mechanism and where it resides.

An example of causal explanation can be drawn from rocky intertidal and subtidal ecosystems. The causal explanation of species richness in communities invokes a complex of mechanisms and contributory processes. In the classical rocky intertidal community with essentially unlimited propagule rain, a particular level of richness is caused by three interacting factors. First is a successional trajectory that is based on a transitive competitive hierarchy, next is the temporal and spatial pattern of disturbance, and third is the resulting patch distributions (Sousa 1985).

It is often possible and useful to cast causal explanations as a nested hierarchy of processes. A phenomenon at a focal level of organization in a hierarchy can be explained by causes at a lower level of organization and by constraints at higher levels of organization (Bartholomew 1982, Ribas et al. 2003, Schoener 1986a, Wimsatt 1984).

An example of hierarchical causal explanation emerges from the commonly observed social behavior of college students at the beginning of an academic year. Multiple causes at several levels of organization may be involved, including needs to show off, to cover insecurities, to prove hormonal assets, and an array of other lowly drives that may combine with a genuine excitement generated by the prospects of acquiring new knowledge. These causes may combine into a more complex intermediate causal mechanism. The puzzling behavior of students in September, readily condemned by those holding the Ph.D., does not generally involve behavioral components of violent crime, damage to laboratories, assaults on instructors, and so on. The apparent self-limitation is not accidental, because it is well constrained by behavior that is traditionally tolerated on campuses. The higher organizational level of academic and societal constraint defines the traditional behavior of students, which permits bubbles of silliness to be vented on occasion. For a more sober example from ecology, to explain assimilation rates when the focal level is the whole plant, one must examine lower hierarchical level phenomena such as individual leaf biochemistry and anatomy, self-shading within the crown, gas fluxes at the leaf surfaces, and stresses the plant has recently undergone. Constraints at higher hierarchical levels of organization or broader scales than the individual leaf include local CO_2 concentrations, shading by other plant crowns, and topographic features, among many others.

In such a nested hierarchy of causal explanation, the familiar phrase "pattern and process" is appropriate. The pattern to be explained resides on a particular hierarchical level, and mechanisms or processes resulting in that pattern are often found at least one level below (Thornley

1980, Passioura 1979). The term "mechanism" in ecology connotes an interaction that is nested within the entity or system to be explained. A mechanism is, therefore, one sort of cause. Other sorts of cause include the enabling and constraining features of the system and the environment in which the system of interest is nested (i.e., at a higher level in the hierarchy). In a landscape-based pattern, such as the spread of one landscape element or patch into another, the embedded mechanisms may be dispersal, establishment, and interaction of organisms, whereas the constraining causes might include herbivore or predator pressure on the potential colonists of the receiving patch. An even higher level cause might be the process creating the patch itself such as fire or wind. Enabling causes might include the general flux of dispersal agents across the landscape, which may be independent of the identity and nature of the specific patches under study. The message is that causes can be a variety of contemporary and historical events and circumstances, can appear on higher and lower hierarchical levels of organization, and can be related to one another in various ways. A cause can itself comprise other more specific and detailed interactions and events (Simon 1973). This composite, hierarchical nature of causes will be explored at greater length when we examine the anatomy of theory. Causation in ecology is best considered to be based on multiple factors and to reside on different interacting levels of organization.

The preceding examples have shown that causal explanation in ecology ranges widely indeed. The hierarchical approach is one way to organize and simplify what could be a confusing mass of observations and relations. Another way to deal with the complexity of causal realms is to divide them into proximal and distal causes (Mayr 1961). Proximal cause refers to all the mechanistic aspects of "how" some ecological phenomenon occurs, whereas distal cause refers to the evolutionary reasons (i.e., basis in natural selection and other evolutionary processes) or historical reasons behind some ecological phenomenon (i.e. "why" the phenomenon occurs). It is worth noting that, evolutionary reasons are, in fact, a subset of historically distal causes that may include such factors as prior land use and history (Williams 1991) or persistent biotic structures or legacies (Chesson 1986, Hannon 1986, Jones et al. 1993, Russell 1993). The important point is that ecological causes often have a significant time dimension. A good example of the distinctions between proximal and distal causal explanation is found in the study of insects on plants. Differences in the degree and plasticity of chemical defenses in fast- versus slow-growing species can be interpreted in terms of contemporary spatial differences in resource availability (Bryant et al. 1983) and evolutionary, habitat-based selection for the same traits (Coley et al. 1985).

3. Testing

Testing is the third component tool for enabling the scientific dialogue between conceptual constructs and observable phenomena. The first two tools can be simplified by saying that generalization is the construction of pattern, and causal explanation is the attachment of a mechanistic meaning to a pattern or phenomenon. In testing, both patterns and explanations can be examined for validity and breadth of application (Box 2.3). Usually, testing examines the assumption(s) used to construct the pattern or process and the logic of the deduction that produces the expectations or reproduces observations to be explained.

All modes of testing share the fundamental similarity of asking whether a stated expectation matches the natural world in a specified domain (Box 2.3). One common form of expectation in the natural sciences is that system x is of a kind to which model y applies (Lloyd 1988). Alternatively, a statement of expectation may posit that a set of circumstances are related to some state or trajectory of a system. For example, a system of interest will change from state x to state z as y increases in intensity. Expectations can refer to different or novel situations or to past,

---

BOX 2.3 Testing and Its Modes

Testing is the comparison of an expectation, or hypothesis derived from theory, against observations from the material world. The expectation need not exist before the phenomena to be used in the test. Several modes of testing exist:

Experiment. Manipulation of a system to generate a reference state or dynamic of known characteristics.
Comparison. Examination of unmanipulated systems to determine their likeness or contrast in state or dynamics.
Correlation. Statistical relationship between measurements of two properties of ecological systems.

---

present, or future conditions. Thus, expectation can be expressed in a variety of ways. All these ways are implied in the shorthand term for a statement of expectation — "prediction." Hence, the literal translation of prediction, "to speak before," is not appropriate for all the kinds of expectations that can be usefully tested. The term "to expect" is used here in its sense of to "look for with some sense of justification." The justification is supplied by conceptual constructs with their empirical base, as we will discuss later (Chapter 3).

Tests of expectations can be of two sorts, either negative or positive. Negative tests fall within the realm of falsification, whereas positive tests fall within the realm of confirmation (Box 2.4). These are complementary approaches to testing. Neither is the sole permissible mode of testing, both can be legitimately used in science, and both have limitations (Amsterdamski 1975, Hill 1985, Miller 1987).

a. Prediction

Prediction is an ingredient of testing and is therefore but one means of evaluating and refining the relationship between conceptual constructs and observable phenomena. Although Peters

---

BOX 2.4 Contrasts between Falsification and Confirmation

Falsification and confirmation, as two methodologies often contrasted philosophically (see Box 1.2), are complementary approaches that fit different research conditions.

Falsification	*Confirmation*
Specific models or hypotheses	General theories
Narrow domain	Broad domain
Tightly designed experiments	Natural patterns or complex experiments
Univariate models	Multicausal models
Logical or necessary outcomes	Probabilistic outcomes
Prone to type I error	Prone to type II error
Single answer	Contingent answers

BOX 2.5 Kinds and Features of Expectation

Prediction. A statement of expectation deduced from the logical structure or derived from the causal structure of a theory.

Forecast. A projection of current trends or conditions into the future; such an expectation may not necessarily be derived from a theory. A statement derived from extrapolation.

Safe Prediction. Those within the confirmed domain of a theory which, if incorrect, would not threaten the basic content or structure of the theory.

Risky Prediction. Those outside the confirmed domain of a theory which would, if incorrect, bring down part or all of a theory; such predictions probe the limits of a theory.

Classification. Expectations of group membership based on similarity or difference in state or dynamics of phenomena.

(1991) argued that prediction should have primacy in ecology, other ecologists (e.g., Shrader-Frechette and McCoy 1994) have pointed out that the use of prediction as the only goal of ecology would be misguided. Still others recognize that prediction, while an important goal, is just one of many (Pace 2001). Irrespective of the importance assigned to this means of testing, and as pointed out earlier, prediction is not independent of understanding. In fact, prediction and its use are valuable to science only because they contribute to generating or revising understanding (Fig. 2.2). One can illustrate this point by a contrasting statement: Successful predictions that do not contribute to understanding are unlikely to be a part of science. For example, predicting that rain will come by reading signs in the entrails of animals may offer great practical benefit, when it works, but it does not contribute to our understanding of the factors causing rain.

Prediction, when broadly defined, can also function as an element of explanation. It does so when it is derived from a theoretical framework to account for already known factual statements or observations. For example, the theory of evolution predicts a shared genetic code among all organisms. Or succession theory predicts that community structure changes in response to the changes in resource availablity at a site through time. More rigorously, we should call such a process postdiction. Postdiction may be more commonly used to evaluate the quality of theories than prior prediction proper, because it is common in ecology that observations precede the development of theoretical constructs.

We have noted already that causal explanation and generalization are two ways to generate predictions, or expectations, about repeatability and scope of pattern, as well as the efficacy and scope of the causes of a phenomenon. Predictions, or justifiable expectations (Box 2.5), can arise in two general ways: deduction from a model and induction from an empirical base. Both deductive and inductive sources of prediction are successful. Examples of each follow.

The search for a universal genetic material was justified by a deduction from the theory of evolution with its tenets of common ancestry and descent with modification. The prediction that trees will possess broad evergreen leaves in tropical wet forests can be considered an induction formulated as the result of the European voyages of exploration in the late 1700s and early 1800s. However, this latter example is complicated by the fact that at least two of the naturalists involved in such voyages, Darwin and Wallace, were clearly theoretically inclined and proposed an evolutionary mechanism from which observations of "convergence" in plant and animal form in similar climates could be deduced. As in this second case, the separation between induction and deduction is most likely always fuzzy. The early search for the C_4 photosynthetic pathway among tropical dry climate grasses may have had a large inductive component. The lists of plants having anatomy associated with C_4 metabolism were accumulated from environments in which the climate was likely to be appropriate, based on past experience. Of course, at the same time, physiologists were developing a theoretical basis explaining the physiological functioning of C_4 metabolism and its ecological significance.

Prediction is a problematical aspect of the philosophy and practice of science because of the forceful claims made about it (e.g., Peters 1980). Prediction is sometimes considered the *sine qua non* of science. Because prediction is both useful and problematical, we must examine it more closely. Recall that there are two stances toward testing, thus there are two stances with respect to prediction as a form of testing: one is falsification and the other is confirmation.

b. Falsification

Falsification was raised to primacy as a criterion of sound science at a particular crisis point in the philosophy of science. The demands that falsification be the only legitimate mode of testing resulted from the logical positivist statement view of theory, with its assumption that laws are literally universal as well as deterministic. If this assumption about the nature of theory is unsupported, then the falsificationist solution loses its force. The crisis in philosophy has been resolved by developments in philosophy itself (Suppe 1977a, 1977b, Boyd et al. 1991), as described in Chapter 1, but dogmatic adherence to falsifiability as a criterion for good science lingers inappropriately in many quarters (Boyd 1991).

The dogmatic application of falsifiability to all aspects of testing is a serious error. The limits to falsifiability are related to the problem of universality and domain, the quality and nature of the statements to be tested, the theoretical context of the predictions, and situations of multiple causality. Demands that falsification be the sole method of testing are based on tacit assumptions that were once considered to apply to all of the scientific enterprise (Hacking 1983) but in fact are only appropriate to certain restricted circumstances. Because falsification can only be used to reject a statement that is thought to be universally true and because statements in most realms of science are not literally universal, falsification can only legitimately be used for statements that should apply uniformly to a specific universe of discourse. Thus, in most areas of science, falsification is likely to be useful for statements of a relatively low level of generality, like the hypotheses appropriate to well-designed, focused experiments or other well-constrained tests. For probabilistic statements appropriate to larger arenas, other modes of testing are likely to be more appropriate (Box 2.3).

Specifying domain is important in framing a legitimate falsifying test. Tests are only appropriate when the domain of the test is known and is congruent with the domain from which the prediction emerged. For example, a test of island biogeography theory that involves certain plants and the insects that can move freely among them is improper because it violates the assumption of "islandness" or isolation of habitat. Because domains — which include process, scale, and hierarchical level, among other criteria — are so rarely well specified in either theoretical or empirical ecology, this kind of error is likely. Recall that falsification is more likely to be appropriate in narrower domains in which the presumed universality of the prediction — a logical necessity for falsification to be a valid strategy — is more likely to exist.

Falsification has other caveats. There is a chance of rejecting a correct explanation. By analogy with statistics, this would be a type I error. For example, concluding that a theory predicting community organization based on unlimited dispersal of propagules is falsified when the theory is applied to a situation in which dispersal is limited is a type I error. Such errors are especially likely to occur in situations in which the proposition subjected to falsification is poorly developed or vaguely cast. Complex compound propositions are sometimes derived from scientific arguments, but because of their compound nature their falsification is difficult to evaluate. For example, the hypothesis that plant stress causes insect outbreaks because of increased nitrogen in leaves (White 1984) is a compound proposition. Plant stress may or may not increase foliar nitrogen. Increased foliar nitrogen may or may not lead to increased insect abundance. A legitimate test would have to segregate the two relationships included in the hypothesis, one of which can be considered a background assumption for the other. Component hypotheses are likely to be used as a shorthand way to give background for a discrete testable statement. Unfortunately, they are confusing and should be avoided. The difficulty of testing compound hypotheses is similar to the difficulty of verification (testing) of models developed for complex open systems. Such systems are not fully isolated from other systems. In such systems, including ecological systems in the field, specification of all the possible influences that could affect the expected result is impossible, and thus this leaves a considerable degree of uncertainty about the outcome of the test (Oreskes et al. 1994).

Even the most apparently straightforward predictions have a theoretical context on which they depend. Auxiliary theories, hypotheses, and assumptions associated with or implied by an expectation are all considered true in even a simple test. The outcome of an attempt at falsification is never more reliable or clearer than the conceptual constructs and assumptions used to frame the test (Fagerström 1987, Pimm 1984). It would be necessary to examine those assumptions before accepting the outcome of the test. Falsification, although it cannot overcome these difficulties any more than verification can (Hill 1985), can be applied cautiously with analysis of the broader context of the predictions (Hilborn and Stearns 1982). Such context includes the background assumptions and the domain of the hypothesis. These should be explicitly examined before the results of an attempted falsification are accepted.

An additional situation in which falsification is sometimes inappropriately applied is in cases of multiple causality. Often, ecologists apply falsification to test whether a cause is relevant to a phenomenon. If a community is influenced by competition, predation, and productivity, for example, a test that controls for only competition may inappropriately reject competition as an influence because its effect is masked by those of the other two processes. A more concrete example is the difficulty of unambiguously falsifying hypotheses condemning "models" of succession of Connell and Slatyer (1977): facilitation, tolerance, and inhibition. The variety of specific mechanisms that contribute to each of these outcomes in the field may make it exceedingly difficult to sort among them (Hils and Vankat 1982). Indeed, various underlying processes interact at any one time in succession (Pickett et al. 1987). When a phenomenon is governed by a variety of processes, falsification of one cause is possible even though that factor is a legitimate part of the causal matrix.

Simple causality has an obvious meaning: one cause maps onto one effect. Well-behaved causality can account for multiple causes, but the outcome of a network of causes can be expressed as a single unambiguous net effect (Jones 1991). Combining causes in specific ways to generate clearly different net effects can sometimes permit falsification to be used in cases of multiple causality. This is especially likely in the case of causal chains. Causal chains involve several linked causes. However, in testing the effects of causal chains, the question is not whether a particular cause operates but what proportion of the effect is due to various causes, what effect is generated by a particular order or combination of causes, or what causes compensate for others. For example, an unusually strong wind may cause an old tree to fall. This will expose the soil previously shaded by the tree canopy to sun. Both the turning over of surface soil by uprooting the tree, and the light, moisture, and humidity in the open spot are altered compared to the previous condition. This change in soil conditions may cause germination of seeds or accelerated growth of seedlings. Finally, the increase in local plant growth in the light gap will cause a modification in the successional trajectory at the site in the immediate proximity of the fallen tree. Seed germination and seedling growth would occur whether the wind was strong or weak, but the growth would occur at different rates. Hence, the wind had only a partial contribution to the determination of the trajectory of succession. In such cases, the ultimate net effect forms the simple, directly falsifiable expectation that a strong gust of wind will change the course of succession.

Falsification is most likely to be appropriately applied to simply put, clear expectations that have an explicit derivation. In this way, the faulty components of the derivation, the incorrect assumptions, the faulty methods, or the problematical empirical basis can be corrected or replaced. Like any powerful tool, falsification must be cautiously applied because of the number of limitations to which it is subject. Not all of these may be evident in advance if a conceptual basis for the prediction is lacking.

Despite the problems with falsification, when it is applied appropriately, it is an extremely powerful tool in ecology. Being able to say definitively how something does not work or behave is a logically compelling conclusion. It can be considered to be "strong inference" (Platt 1964) because, in one fell swoop, a conclusion can be reached. It guards against a type II error of erroneously accepting an incorrect hypothesis. Falsification can be effective, however, only if the problems enumerated earlier do not occur. Hence, it works best in cases in which (1) the expectations to be put to test are well developed (that is, they are explicitly derived from a conceptual base); (2) the expectations contain well-defined and measurable components whose modes of interaction are unequivocal; (3) the domain of the test is appropriately narrow so that the assumption of universality can be met, and such a domain is precisely congruent with the domain of the theoretical structure from which the expectation is derived; and (4) causality is simple or well behaved. It is worth emphasizing again that philosophers recognize that falsification is most seriously limited by the inability to know or control for all the relevant auxiliary hypotheses that can affect the outcome of a test (Boyd 1991) and by its legitimacy only in testing supposedly universal statements (Hull 1988).

c. Confirmation

The second stance toward predictions or statements of expectation is confirmation. Confirmation is especially effective in guarding against the rejection of a correct explanation to which falsification is prone. In statistical analysis, rejecting a correct hypothesis is labeled a type I error. The opposite kind of error, accepting a false hypothesis, is a type II error. Because confirmation is susceptible to type II errors of accepting an incorrect explanation, it is best viewed as a suite of methods involving considerable redundancy and cross-checks. In other words, confirmation is a valid scientific approach because it is actually a composite method. Confirmation is most often used for probabilistic relationships, compound conceptual systems, multicausal systems, or whole models or theories, in contrast to the focus of falsification on univariate, universal expectations.

The suite of methods that constitute confirmation has been analyzed philosophically by Lloyd (1988) using evolutionary theory. We find her analysis to be applicable to ecology as well. In general terms, successful confirmation consists of three components: (1) the degree of fit between data and theory, (2) the independent support of assumptions of the theory, and (3) the variety of independent evidence (Box 2.6). We explain each of these components of confirmation more fully.

Fit is determined by congruence of the patterns observed in nature with the patterns predicted by the theory to be tested. To put this another way, fit is the finding that the expectations of the theory are met in nature. Processes, interactions, relationships, and outcomes are examples of predictions that can be used to determine fit. Note that the degree of fit can be assessed by particular and focused attempts at falsification.

BOX 2.6 Modes of Confirmation (Based on Lloyd 1988)

Confirmation of theory consists of confirming empirical claims made by the models of the theory. Evidence confirms a claim if it gives additional reason to accept the claim. The issue in confirmation is whether a particular system or kind of system is isomorphic with certain aspects of a model. That is, do the system and the model conform to one another. Because models necessarily simplify nature, isomorphy between the observed, material world and models cannot be complete. The modes of confirmation are these:

Fit between Model and Data. Statistical comparison between the material world and a model. This may involve defining constants, curve fitting, and experimental outcomes.

Independent Testing of Aspects of the Model. Independence refers to origins separate from outside of the domain of the target theory. Many kinds of independent tests exist, including separate tests or confirmation of the empirical assumptions of the model; confirmation of the structural or analytical appropriateness of the assumptions; and appropriateness of relationships specified in the model.

Variety of Evidence. Variety of instances in which model output matches the material world: range (magnitude, scale) of instances over which model output matches the material world; variety and number of assumptions tested independently; and types of evidence available.

Examples of fit are many. A forest simulation model might demand a particular species composition at a certain altitude on a mountain range (Shugart 1989). The degree of fit could be compared using multivariate statistics or indices of community similarity. Likewise, a population genetics model might report a gene frequency value that can be compared with a natural population the model is supposed to represent (Lloyd 1988). Alternatively, a model of population regulation might produce a graph of population density through time that could be compared statistically with the trace produced by a target natural population. The statistical null hypothesis in each of these cases is a focused prediction subject to falsification.

The second way to evaluate confirmation of a theory is via the independent support of assumptions. If the conceptual construct being tested has as its foundation a large number of assumptions or other components that are themselves well confirmed or have stood the test of attempted falsification, then that theory has a high degree of confirmation. Finding major expectations from such a well-confirmed theory to be falsified would more appropriately lead to an examination of the parameters of the test rather than to the immediate rejection of the theory.

An example of independent support of a conceptual construct appears in gradient theory. Gradient theory in ecology is based on the idea that ordered environments underlie differential responses of ecological systems. The assumption that gradient theory is generally applicable is shown by the fact that many natural environments are found to be heterogeneous and that the heterogeneity is orderable in terms of the major factors producing the differences. Thus, real concrete gradients appear, for example, on hillsides and in the rocky intertidal zone. Abstracted gradients are found using various statistical techniques. These techniques quantitatively order variation that is dispersed in space.

The assumption that ecological entities respond differentially along gradients has also been examined in many systems using many techniques (Austin 1985, Vitousek and Matson 1991,

Whittaker 1975). This aspect of gradient theory appears as dose-response curves of various populations, of community characteristics such as diversity and density, and of ecosystem features such as productivity and nutrient dynamics. Abstract ordering of environments through time, as successional gradients, exposes biotic response curves as well. Responses of species, population, community, and ecosystem properties have also been generated experimentally using natural, anthropogenic, and constructed gradients. Finally, the assumption of a mechanistic link between the environmental pattern and the ecological responses is confirmed by knowledge from physiological ecology and from models that explain population and community responses along gradients by relying on individual performance (Huston et al. 1988). Thus, gradient theory ties many fundamental features of organismal, community, and geographic ecology together. It relies on a large number of assumptions that are themselves the accepted outcomes of different theories. These assumptions were developed separately and independently of gradient theory.

The final criterion for confirmation is variety of evidence. No single instance of good fit will be sufficient to confirm a theory. Gradient theory well illustrates how variety of evidence is used in confirmation. Fit between the gradient patterns expected and those actually found in nature has been examined repeatedly. The way fit is examined ranges from direct observation, to ordination and modeling, to experiment. The responses of organisms and ecological systems have a theoretical and mechanistic basis and have likewise been examined in many situations using different empirical and abstract methods. Good fit of gradient theory has been found in terrestrial and aquatic systems, freshwater and marine realms, plant and animal ecology, short- and long-term studies, and a wide range of geographical settings from tropics to polar regions. These are but a few examples of the variety of evidence relevant to a general and fundamental ecological theory.

Overall, confirmation of a theory requires many different cases of fit between its propositions and reality, as well as a number of independently supported assumptions (Box 2.6). Confirmation does not depend on the inclusion of one accepted assumption or model component in a theory. Unlike falsification, for which one sound negative instance of a purported universal relationship is sufficient, confirmation requires a variety of evidence; no simple a priori rule exists of how good the fit must be, nor how many instances of fit there must be, nor how many independently supported assumptions there must be, nor how diverse the evidence must be in general. Like the decisions implied for confirmation by the preceding list, falsification also entails the need to make decisions, but the decisions for falsification relate to certainty about the domain of the prediction and about the completeness, soundness, and relevance of the auxiliary hypotheses surrounding the test.

Confirmation does have some significant advantages for ecology. Because it deals well with complex conceptual constructs, for example, models and whole theories, it is especially appropriate for evaluating multiple causality. Because ecology is a relatively young science and its subject matter is highly contingent and multicausal (Gould 1989, Hilborn and Stearns 1982), confirmation may be an especially appropriate tool for ecologists. Many other examples exist of the multiplicity of methods used in confirmation. Darwin's use of a variety of deductive and inductive methods in generating and supporting his theory is a model for confirmation in biological sciences (Ghiselin 1969, Gould 1984, Grene 1985, Lloyd 1983, 1987, May 1981, McIntosh 1987). Levins's (1966) suggestion that multiple models be aimed at a problem is another example of the multiplicity of confirmation. Vermeij (1987) summarized a similar view of the methods for discovering mechanisms in the historical and evolutionary sciences, including (1) constructing complementary possible explanations, (2) rejecting only those that are clearly wrong, (3) discriminating among mechanisms based on the amount of support each gains from independent lines of evidence, and (4) discriminating among mechanisms based on soundness of theory underlying each one. These diverse approaches to confirmation share many features and are contemporary examples of the concept of "consilience of inference" (Whewell, in Ruse 1979). Whewell was perhaps the foremost philosopher of science in Darwin's time, and his methodology relies on multiple lines of evidence converging on a central model.

The relative youth of ecology as a science suggests that many patterns are yet to be discerned from nature. Accumulating and evaluating the generality of patterns by quantitative, statistical, and inductive processes are still a major need in ecology (Maurer 1999). Much care, creativity, and quantitative and statistical rigor can be brought to bear on this task. The fact that ecological systems and processes are almost universally highly contingent on prior states of the site or system and so often subject to multiple causality suggests that confirmation as a strategy will be highly productive. This conclusion is reinforced by the observation that the dynamics or condition of a focal system often depends on the states and dynamics of adjacent or even distant systems. The question of how to divide the world into cases of similarly behaving or similarly structured populations, communities, ecosystems, and landscapes is well served by confirmation and is a major challenge in ecology.

Confirmation is an intentionally constructive process, but one that has checks and balances. Confirmation should be especially useful in constructing new models and theories to permit integration in ecology. Falsification is intentionally a destructive process. If one has a highly integrated and explicit set of conceptual constructs and truly universal predictions, then falsification can be used to excise faulty parts or even to destroy the entire edifice if it proves to be fundamentally at odds with nature. Ironically, multiple instances of failed falsification will confer soundness and legitimacy to conceptual constructs summarizing and explaining a pattern or phenomenon, and by so doing they increase the degree of confirmation. Thus, confirmation and falsification are complementary processes; each has an important role to play in improving integration in ecology (Box 2.4).

D. Understanding and Application

Ecological management is the application of scientific expertise from ecology to a societally defined problem. The definition of problems can arise from individuals, communities, private institutions, or public agencies. Scientific understanding interacts with management via three tools: forecast, prediction, and classification (Fig. 2.2). We will exemplify in this section how these three tools for application relate to understanding.

Forecast is the straightforward projection of current states into the future (Box 2.5; Lehman 1986, Mayr 1961, Pielou 1981). This is simple extrapolation. Such projections are often called for in management. Will the amount of a toxic agent increase in the future? Given current release rates of the toxin into the environment, we can forecast that the total amount in the environment will increase. Will diversity of a polluted lake decline? Given known impacts of pollutants on lake biota, we can forecast that diversity will decline. In both cases, we can make these statements without recourse to any additional information. Tacitly then, forecasting assumes that the same boundary conditions and mechanisms that have held for a system in the *past* will continue to hold in the future, even if those conditions and mechanisms are unknown. This is clearly a dangerous assumption, because it is often made in ignorance and is therefore impossible to evaluate. For example, toxin release rates may decline because of changes in manufacturing processes, or the toxin may be degraded in the environment at higher rates than it is added. The aforementioned lake may contain biota that are resistant to the pollutant because of the history of pollution and selection, so that no further decline in diversity occurs.

Forecasts are made as extrapolations of current trends or are frequency or probability distributions of some phenomenon. As an example of forecasting by probabilities, the relative distribution of seedlings beneath canopy trees can be used to forecast the species identity of the new tree that might dominate the canopy after gap formation and filling (Horn 1971). Similarly, the recurrence of natural events, such as floods or earthquakes, can be used to forecast future occurrences by specifying a probability of recurrence per specified time interval. The 100-year flood and an outbreak of locusts are examples. Forecasts of this sort need not rest on strong scientific understanding. Rather, they may be founded primarily on a collection of relatively undigested empirical observations. Of course, there is value in forecasts, especially when dealing with areas in which there are no or only rudimentary explanatory models or in which the existing models are not at the appropriate scale.

Despite the fact that a forecast may not arise from a deep scientific understanding, it may feed back on the development of theory and, hence, of integration. When a straightforward forecast fails, it may suggest the need for a new model, causal explanation, or entire theory. More specifically, the way in which a forecast fails may suggest the form of the causal explanation that may be appropriate for a phenomenon. For example, one of us made a forecast, based on numbers of graduate students entering ecology programs in the 1960s, that in 2007 all the humans on Earth would be ecology Ph.D.s. The 1980s and 1990s showed a progressively slower growth of the field of ecology, akin to the logistical population growth curve, suggesting an increasing habitat resistance model. Had this growth stopped suddenly, an alternative explanation would have been needed involving a catastrophic change of behavior (e.g., a mass migration to Wall Street). Finally, forecast and theory can have a more complex relationship when they blend in various ways. In a study of effects of an exotic species, zebra mussel, invading the Hudson River, a mechanistic model was superimposed on a time series forecast in order to generate projections of rotifer abundance and to compare those projections with yet another observed time series (Pace 2001).

Prediction, as a statement of expectation based on some explicit justification, is the second and stronger link between management and theory (Box 2.5). Predictions, in the strict sense we defined earlier, are necessarily generated by theories. In other words, theory provides the explicit justification by articulating the logical and empirical reasons for holding a particular expectation are well articulated. When such a prediction fails in a management context, the reason should be suggested by the structure and content of the theory from which the prediction was derived. The failure, if the parameters of the management case in fact match the scope of the theory, can call attention to the faulty or limited aspects of that theory. Failure of a prediction derived from theory in a management context can suggest whether the whole theory needs to be replaced, whether some part of the theory must be replaced, or whether a new theory is needed. Failure to restore oak savannas on the prairie margin using the same techniques and species that had worked on the true prairies pointed out a weakness in the theory of plant community organization and structure in that region (Jordan 1993). It turned out that the savannas were not simply a mixture of prairie and forest species but a somewhat distinct formation.

While it is important to discriminate, as we have done here, between forecast and prediction as distinctly different tools for interaction of understanding with management, these two tools can be confused in common, everyday use. While our strict definition of forecast is essentially equivalent to extrapolation, many commonly recognized forecasts have theoretical components. Even weather forecasts now have an impressive theoretical component, as illustrated by the many models used to generate the nightly weather spot. Prediction, which we have defined as an expectation derived from a theoretical base, may for practical purposes often have probabilistic, descriptive components. For example, one can "forecast" the spread of Africanized bees based on past rates of spread, but the "can" part arises from the knowledge of population dispersal imperatives and ecological tolerance of the species. In other words, both a theoretical and an empirical base is used in making the so-called forecast of Africanized bee migration. We will hold to the distinction between forecast and prediction in this book.

Classification is the third and final tool. Classification is the division of a universe of discourse or topic of interest into appropriate portions. Societal problems often require classifications. For example, what constitutes a wetland? What are the rates at which various invasive species will respond to manipulation? What is an indicator species for a certain kind of pollution? What constitutes an endangered species? Such classifications can suggest the need for ecology to consider a problem more deeply. For example, conservation concerns pointed to a refined view of rarity in which several previously unappreciated causes of rarity were discovered (Fiedler and Ahouse 1992, Gaston 1994). Fiedler and Ahouse (1992) contrasted rare species in a space, defined by axes of wide versus narrow distribution and short versus long persistence. Likewise, contrasting management-driven classifications of forests in different regions — for example, between the "virgin" forests of North America and the "ancient woodlands" of Britain — exposed hidden assumptions about the structure and dynamics of ecological communities. The admired ancient woodlands of Britain often incorporate and rely on important direct human uses of timber, wood products, and grazing, whereas the old growth forests of North America rely on a different mixture of human and natural influences (Williams 1991).

E. Kinds of Ecology

The outline of understanding and its components, and their relationship to management, suggests a broad overview of the various approaches that can be taken toward ecology. There are different ways to divide understanding and group its component features and tools. Basic ecology encompasses both empirical and theoretical specialties. In basic ecology, emphasis is on the observables and the conceptual constructs and on the tools and statement of domain that link the conceptual constructs and observable phenomena. Examples of *observables* in basic ecology include light flux, individual organisms and compilations of their numbers or biomass over time, seed dispersal distances, frequency distributions of body sizes, plant cover, and nutrient concentrations, among others. Examples of *conceptual constructs* include light compensation point, the guild concept, trophic level, element cycling, or island biogeography theory. Conceptual constructs can thus refer to processes such as growth models, properties of models such as stochastic density vagueness, or complex ideas such as density dependence mediated by extrinsic factors. Examples of *tools* include methods such as the light and dark bottle procedure to measure algal production, protocols for data collection such as life tables, modes for translating population parameters into abstract models of dynamics, and conventions such as statistical standards. Although a practitioner of basic ecology will likely not tackle all this richness alone, the whole community of basic ecologists must effectively cover this broad territory and stimulate communication among those who specialize in various methods and approaches. Within basic ecology, we can identify several notable specialties. Theoretical ecology as a sort of basic ecology focuses on the generation, refinement, and derivation of expectations from conceptual constructs. Likewise, empirical ecologists focus on the discovery, documentation, and generalization of the phenomena of ecology. Importantly, however, both basic empirical ecology and theoretical ecology overlap in their concern with the interactions between the conceptual constructs and the phenomena. Theoretical and empirical ecologists must be and are concerned with the tools and specification of domain for an ecological subject. Thus, empiricists and theoreticians propose, execute, and evaluate causal explanations, generalizations, and tests, as well as deal with objects, relationships, dynamics, and scale. The overlaps are substantial; integration in ecology can be well served by emphasizing these overlaps between theoretical and empirical approaches rather than their separations. Indeed, modern philosophy recognizes such a creative intermingling (Hacking 1983) in contrast to classical philosophy, which took pains to separate the empirical and theoretical aspects of science (e.g., Carnap, 1966).

Applied ecology deals primarily with the use of ecology in societally mandated tasks, including such specific activities as conservation, restoration, and resource management. Although the extraction of the relevant portions of ecological understanding for management concerns might seem to leave little of ecology within the scope of applied ecology, the reverse is in fact true. Management may provide extremely useful tests of basic ecological understanding. Is basic ecological understanding complete? If it can successfully deal with the often novel situations generated by societal problems, then it is likely to be complete. Are the models and theories sound and well developed? If they can make adequate predictions and forecasts in appropriate domains when funds and time are on the line, then they are likely to be good models. Is it clear which models apply to specific problems and situations? If so, then the classifications contributing to ecological understanding are well developed and extensive.

In showing how the different kinds of practice in applied and basic ecology, and within basic ecology, combine empirical and theoretical approaches, it emerges that ecology is not cleanly divisible among these approaches. In fact, the connections between the various aspects of ecology — including explanation, generalization, testing, specification of objects, determination of relationships and dynamics, and the provision of classifications, forecasts, and predictions — hold the discipline together. These tools and connections require contact between practitioners of the various kinds of ecology and their subject matter. Because ecology is the study of relationships, it is most appropriate that it be glued firmly together by the relationships between concepts, phenomena, and management.

IV. Conclusions and Prospects

This chapter has examined the "arrows" or processes connecting the "boxes" or parts that constitute understanding (Fig. 2.2). This examination has been motivated by the promise that greater integration can result from paying explicit attention to how the parts of ecology can be used together. To further our comprehension of how to enhance integration in ecology, we must now look at the poles of the dialogue that constitute understanding. Chapter 3 examines the anatomy of theory. We will not look at management in more detail, but we will leave that to individuals who have a better command of the examples, difficulties, and successes in that field (e.g., Biggs and Rogers 2003, Christensen et al. 1996, Fiedler and Jain 1992, Luken 1990, Starfield and Bleloch 1986). Neither will we look further at the observable phenomena, since the empirical side of ecological understanding is especially readily available in the general and advanced textbooks of ecology (e.g., Begon et al. 1996, Fitter and Hay 1987, Gurevitch et al. 2006, Schlesinger 1991, Stearns 1992). The structure and use of theory, however, have been most often examined in disciplines other than ecology, usually physics and less, but increasingly often, in evolution. Because ecological integration can be enhanced by a more explicit attention to generating and evaluating understanding, the structure and use of theory must be analyzed in ecological terms. That is the central purpose of this book.

Part II

The Nature of Theory

3

The Anatomy of Theory

"Every map is a simplification of a real landscape; nevertheless, maps are enormously helpful, and it is hard to imagine how we could get along without them."
Raymo 1991:147

I. Overview

Increasing integration in ecology is the motivation for our analysis of the anatomy, or structure, of theory. Understanding is the state by which science achieves and evaluates integration. Understanding requires conceptual constructs, a specified universe of observable phenomena, and the tools to permit dialogue between them. We use the general word, "phenomena," to reflect the wide range of things that science studies: things, events, interactions, and so on. Theory is thus one of the pillars of scientific understanding. But if theory is to be most useful in advancing understanding and promoting integration, exactly what it is and how it functions and changes must be known. An additional limitation to the use of theory as a major integrative tool in ecology is the narrow view that many ecologists have of what constitutes theory. Here, we will begin with the broadest, most inclusive definition of theory, and we will discuss the various components and subtypes of theory. Theory is the touchstone of understanding. This chapter lays out the nature of theory and suggests how it is constructed and delimited. We will define the components of theory, point out their distinctions from one another, and give examples from well-known ecological theories and from evolutionary theory. We will present the components of theory in roughly the order of increasing complexity, degree of derivation, or dependence on other components of the conceptual system.

II. Theory and Its Conceptual Foundation

Theory is perhaps the most important tool for integration in ecology. Without knowledge of the breadth and content of theory, important functions of theory may be neglected. This neglect can compromise the dialogue between conceptual and observable phenomena. Precision in recognizing the parts of theory is also important for evaluating the status of a theory and therefore the role it can play in understanding at any given stage of its development. Finally, theory can improve a scientist's performance in less tangible ways (Box 3.1). But all these values of theory are difficult to comprehend if theory is only vaguely appreciated or is thought to be restricted to

61

BOX 3.1 Mostly Serious Rationales for Sensitivity to Theory in Ecology

Gets you through days when the instruments are not working, the organisms are not
    cooperating, or the weather is too awful to go out in the field
Prevents you from getting lost in the threatening tide of details
Lifts you out of the suction of the reductionist downward spiral
Helps you make decisions about what to do next in a world in which everything is a little
    bit interesting but only some things are truly worthwhile
Provides a framework on which to reassemble all the disparate threads of your
    dissertation
Gives satisfaction in identifying situations in which arguments take place in different
    domains

---

mathematical models. Theory may seem to be a mysterious commodity, since it is so commonly
mentioned, so widely admired, and so variously defined. This chapter aims to bring some order
and clarity to the complexities surrounding this fundamental component of the scientific
enterprise.

A. Definition of Theory

Theory is a system of conceptual constructs (Suppe 1977a). This definition implies two basic
aspects of theory. First, because theories are conceptual *constructs*, they are composed of various
parts or specific components. Second, because theories are *systems* of such parts, their components must have some order and must *interact* via combination, derivation, inference, entailment,
or other logical or empirical relationships. These relationships will be clarified in discussing each
of the individual components of theory (Box 3.2). It is important to emphasize that any theoretical system has a specific domain, and that it affords causal explanation of observable phenomena within that domain (Miller 1987). Recognition that theories are systems of conceptual
constructs (formalisms and nonobservables) tied to an empirical base (measurables and observables) is indicative of the view of theory called scientific realism (Scheiner 1994). These features
of the definition indicate that theory performs a particular job in science and has a specific realm
in which it applies. These features can be reiterated in the broadest and most comprehensive
definition of theory: *A theory is a system of conceptual constructs that organizes and explains the
observable phenomena in a stated domain of interest* (Box 3.3).

   Although we will define each of the components of theory as precisely as possible, we must
caution that individual components only have meaning as part of the overall system. Like the
words of a spoken language, the components of theory have specific meanings, but only in the
context of the rest of the language and its relation to the empirical realm.

   In addition, their relationships to other components of a theory can be modified as the theory
changes or as its dialogue with observable phenomena develops. Hence, a theory rests to a significant extent on changeable relationships about the natural world, which means it is malleable.
This ability to change is addressed more fully in the next chapter. Again, the malleability of
components of theory is like the malleability of words that shift meaning as the needs and circumstances of a language change. New words are invented to deal with new inventions, new

---

BOX 3.2 Components of Theory

Domain. The scope in space, time, and phenomena addressed by a theory; specification of the universe of discourse for a theory

Assumptions. Conditions or structures needed to build the theory

Concepts. Labeled regularities in phenomena

Definitions. Conventions and prescriptions necessary for the theory to work with clarity

Facts. Confirmable records of phenomena

Confirmed generalizations. Condensations and abstractions from a body of facts that have been tested or systematically observed

Laws. Conditional statements of relationship or causation, statements of identity, or statements of process that hold within a universe of discourse

Models. Conceptual constructs that represent or simplify the structure and interactions in the material world

Translation modes. Procedures and concepts needed to move from the abstractions of a theory to the specifics of application or test or vice versa

Hypotheses. Testable statements derived from or representing various components of theory

Framework. Nested causal or logical structure of a theory

---

situations, and contact with different cultures. A further complication with conceptual constructs within a theory is that they can have different degrees of complexity. They may be simple, straightforward concepts that refer to a relatively narrow idea. Or they may be compounded from other, simpler components of a theory. In such a case, conceptual constructs may be derived from one or more other components of theory. The derived nature of complex components of theory is yet another reason for their dependence on other components for both their meaning and their use.

BOX 3.3 More on the Definition of Theory

The concise definition that *a theory is a system of conceptual constructs that organizes and explains the observable phenomena in a stated domain of interest* captures most if not all the attributes of theory accepted by the modern view. By viewing theory as a system, one implies a degree of coherence and interdependence of components embedded in the framework. Note that components (Box 3.2) include facts, generalizations of observed phenomena, procedures used to relate more abstract components to direct or indirect observations, and others, as well as the specification of the domain. All of these components jointly define methodology — they help to identify what kinds of data are legitimate, how to collect them, how the data inform the theory, what the rules of testing are, when the observations become established facts, and when the theory shows inconsistencies in need of further work. Finally, this definition of theory implicitly makes it a human enterprise because its content changes as facts accumulate and conceptual structure becomes refined.

III. The Basic Conceptual Content of Theory

We present the components of theories and their precursors in three groups: the basic conceptual content, the empirical content, and the derived conceptual content. No theory can operate without components that play each of these kinds of role.

A. Pretheoretic Notions

Strictly speaking, notions are not parts of theories, because theories are a device to make assumptions and conceptual structures explicit. This is why theories make the structures of scientific arguments clear and usable. Notions do not articulate assumptions and do not have an explicit conceptual structure. However, because notions are often the *raw material* from which theory is generated, we include that idea for completeness. Notions are metaphors, analogies, visual pictures, personal intuitions, or vague guesses about how the world is or behaves. Notions are closely related to the flashes of insight that identify novel problems or novel solutions to a problem and, therefore, are pretheoretic. Notions, in contrast to the clear structure of theories, are usually imprecise and may be tentatively or incompletely articulated. As a theory develops and becomes more complete, notions are replaced by concrete components of theory whose structure, rationales, and implications are made explicit, communicable, and analyzable by the scientific community. Examples of notions include Kekule's dream about snakes eating their own tails, which presaged the structure of the benzene ring, or Clements's notion of succession arising from his youthful observations that the disturbance of westward bound wagon trains left impressions on the prairie that gradually disappeared. Later, Clements also employed another notion, a view of ecosystems as superorganisms. That notion was essentially undeveloped in Clements's theory and so could never be used or unambiguously evaluated. The problems that have plagued superorganism views in ecology suggest the need to prevent confusing a notion with a theory. "Balance of nature" was another notion, primarily associated with a 19th century European worldview, that inspired development of research into the stability of ecological systems and the subsequent crystalization of its assumptions, logic, and tests (Cuddington 2001, Egerton 1973; Odenbaugh 2001). However, this is a work in progress and terms such as "stability," "integrity," and "fragility" remain still somewhat vague (Pimm 1991). Metaphors have immense power to suggest novel interpretations, ideas, and syntheses, but they are not in and of themselves theories (Pickett and Cadenasso 2002).

B. Assumptions

Assumptions are the explicit presumptions about the nature of the system of interest. Assumptions state what components and interactions will be included, the structure of the models to be employed to represent the system of observed phenomena, the facts that will be accepted into the theory, and the initial or external conditions for the system to exist or behave in a certain way. In other words, assumptions are the conditions needed to justify the content and structure of the theory (Lewis 1982, Lloyd 1987, Murray 1986, Stegmüller 1976). In a poorly developed theory, some assumptions may be implicit rather than clearly stated.

Assumptions can take various forms. They can appear (1) as postulates — that is, conventions about the meaning of terms or the nature of relationships; (2) as boundary conditions; (3) as facts accepted from some other theory; or (4) as relationships between such facts (Lewis 1982). An important feature of assumptions is that often they involve a certain amount of intelligent guessing or choosing from plausible alternatives. Discussions that proceed without clear statement of the assumptions that are made will likely be problematic because the participants may not in fact be talking about the same system. If assumptions are not stated, each person involved in a discussion may well assume a particular structure and dynamic that differ from that assumed by the other participants. So if two people talk about apples and oranges without stating what they mean, they will likely have a fruitless discussion!

We can give examples of various kinds of ecological assumptions. First, two common boundary conditions exist in the classical theory of temperate climate terrestrial plant succession. The smallest time step is a year, so community dynamics on the seasonal scale are not considered successional in the classical theory. The upper time limit is on the order of centuries, so natural climatic changes are usually excluded. In our second example, evolution, facts are absorbed from other disciplines as assumptions. Geology, for example, provided the stratigraphy from which changes in lineages were inferred. This geological insight was taken into evolution under the label of "the law of superposition." Here, evolutionary biologists accept the methodology of geology when interpreting upper depositional strata as being younger and thus containing more recent organisms. Another evolutionary assumption is the nature of the gene. The fact of the gene was ultimately absorbed into evolutionary theory, but only after refinement by other disciplines. The contemporary idea of the nature of the gene was provided by molecular biology after a long period of development with contributions from various other scientific perspectives (Darden 1991). Finally, a relationship adopted from physiological ecology into biogeography is based on the $Q10$, or increase in metabolic rate with a temperature increase of 10 degrees C. The $Q10$ of biotic processes is >2, which indicates a process requiring biogeographic explanation, compared with physical processes, which have a $Q10$ of ~1. Physically driven changes in metabolic rate do not require additional biogeographic information for their explanation. Each of these assumptions for one theory is a fact or generalization in another theory.

An axiom, or self-evident proposition, is a special case of assumption. The best-known examples of axioms, those from geometry, are unlikely to have parallels in most of natural science because geometric axioms lack empirical content. For example, various radically different geometries (e.g., Euclidian, Riemannian) can be established by relying on different axioms, none of which have to reflect any particular body of empirical fact. Axioms can exist in highly abstract or conceptually motivated ecological theories, but we do not expect them to be common in ecology. Ford (2000) presented many empirical generalizations as axioms (e.g., pages 108–122); however, this is inconsistent with the definition of axiom we presented earlier. The informal use of self-evident propositions, such as the statement "all mammals are animals," has substantial empirical content and is, again, not axiomatic within the above definition. While these two examples are not, from our point of view, axiomatic, it is important to recognize that such empirical generalizations may serve as assumptions.

Assumptions that have empirical content (i.e., are nonaxiomatic) may be subject to direct test under certain conditions, especially when they are directly addressed by models or as hypotheses that emerge from the theory. For example, one of the assumptions of succession theory is the lack of a directional climate change. This assumption is reasonable at the scale of plant replacement normally considered by ecologists; and it is necessary for the interpretation of the successional models, but it is not immutable. Indeed, human accelerated climate change would likely violate this assumption and would require revision of succession theory. Alternatively, structural assumptions, those that are embodied in how models or other conceptual constructs are built, can be evaluated only via the effectiveness of the conceptual device they underwrite. Thus, structural assumptions are tested only indirectly as models are verified, compared, and developed, or when the theory of which they are an integral part is rejected or confirmed (Lloyd 1988, Murray 1986). For example, structural assumptions in models include the use of difference rather than differential equations. Differential equations assume instantaneous interactions, as opposed to difference equations, which specify a time step for interactions. In many ecological systems, history and time lags affect interactions, so that differential equations may yield faulty predictions. Another instance of structural assumptions comes from the theory of island biogeography. For the theory to be applied, a particular temporal and spatial scale is assumed. The scale must be small enough that occasional transfers of individuals take place but large enough that they do not influence dynamics of the island population. Any shift of scale away from the one prescribed earlier means a switch to either metapopulation dynamics or evolutionary processes.

We can cite several further examples of assumptions in ecological and evolutionary theories. Evolutionary biology assumes that individuals are distinguishable and distinct. Density is nothing more than an enumeration of individuals. The purpose of measuring density is the assumption that its value represents the integration of birth, death, immigration, and emigration. As a first approximation, all individuals are considered the same. For certain models, differences among individuals, such as between juveniles and adults or between males and females, are taken into account. Because evolution is a pattern of descent with modification or the alteration of heritable differences over time, such assumptions about the nature of individuals are a crucial part of the theory.

Other kinds of ecology also involve assumptions about how their focal entities enter into interactions. In community ecology, spatial proximity is assumed to be necessary for reciprocal resource-based interactions. In succession theory, species from contrasting portions of a successional series are assumed to differ in biological characteristics (Pickett 1976, Tilman 1988). In ecosystem ecology, all ecosystems are assumed to be bounded for the purposes of the mass balance constraint used in determining nutrient budgets with their inputs and outputs (Likens 1992). In island biogeography theory, equilibrium between extinction and immigration is assumed to emerge and persist, given sufficient time (MacArthur and Wilson 1967). These examples of assumptions show how fundamental and pervasive they are.

C. Concepts

Concepts are defined as regularities in events or objects designated by a label (Novak and Gowin 1984). They may take many forms and have a number of characteristics. For example, concepts can be simple or compound. They are usually broader and more abstract than the particular instances of events of objects that they encompass. Concepts are constructed from many observations, so they represent an abstraction of the regularity from these observations. Concepts can refer to individual objects or to classes of phenomena or relationships (Leary 1985). We will explore this richness of characteristics and give examples later.

First, it is necessary to differentiate concepts both from the pretheoretic notions we have encountered as the seeds of theory and from the more highly derived components of theory. On the one hand, concepts differ from vague notions because they are not subjective but are explicit and can be communicated and evaluated by a community of researchers. It is also important to note that concepts that are well developed in one area can become part of other constructs elsewhere in science. Simple concepts differ from, for example, compound concepts, highly derived models, and complete theories because simple concepts are not built from other concepts, do not involve a high degree of abstraction or idealization, do not have internal logical structure, and do not themselves generate empirical implications. Now armed with an appreciation of some of the key features of concepts, we can explore some ecological examples.

Because there are many kinds of concepts, we begin with simple examples of ecological concepts and move to more complex ones. An example of a simple concept is that of "tree." Abstracted from numerous observations of plants that have a central, persistent, woody support structure dividing into smaller supports that bear leaves, the concept of "tree" differs from the

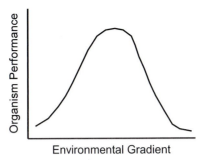

Figure 3.1   Adaptation as a continuum. Like most concepts relevant to ecology, adaptation as a state can be present to various degrees. Along some environmental gradient, organisms will exhibit differing degrees of conformity to the variation in the environment through differences in their performance or survival.

oak tree just outside the window. In addition, even though oak and maple trees differ from one another in many important ways, they share the common features of the concept of "tree." More complex concepts are also used in ecology. As an example of a complex concept, "adaptation" embodies the idea that environments differ in a way that can be ordered and that organisms differ in the degree to which their structure or function matches the range of environments available. By analogy to the tree example, specific environments differ among each other, organisms differ in traits allowing them to live in those environments, and the degree to which an organism matches the environment can also vary greatly among organisms-environment pairs. This complex concept can be captured in the statement that adaptation is the degree to which organisms conform to environment (Fig. 3.1). Like many concepts relevant to ecology, adaptation is best considered as a continuum. Organisms can conform, or fit, better or worse to particular environments. Note the relevance of these ideas to the outline of gradient theory given in Chapter 2.

Competition is another complex ecological concept. The fundamental idea behind this concept is that many resources are limited in their availability, more than one organism may use the same resource, and therefore there is potential for joint use of the same limiting resource. The most fundamental way to view competition is, therefore, the *process* of concurrent use of a limiting resource. The term "competition" is, however, often used in a somewhat different but closely related way as well. Given the previously stated fundamental requirement for concurrent use of a limiting resource, then if the organisms differ in their capacity to use this same resource, the organism with the greater capacity for resource use will obtain a greater fraction of the resource, while the other organism will obtain less of this limiting resource. Assuming some effect of the amount of resource on the performance of the organism (e.g., growth or reproduction), this differential usage capacity results in greater performance of one organism (the "superior competitor") compared to the other (the "inferior competitor"). As a consequence, "competition" is often used to refer to the *effect or outcome* of the competition process. Sometimes, and more loosely and inappropriately, the term is also used as a general descriptor of any negative interaction among organisms. However, since a negative outcome can clearly arise via processes other than competition, using the term "competition" to describe negative outcomes when the process has not been demonstrated is not scientifically defensible. Neither of the two usages presented earlier should be confused with the underlying *mechanisms* by which organisms affect the amount of a limiting resource available to other organisms (e.g., through foraging ability, physiological processes of plant water use, or plant architectural features affecting light capture) that then result in differential performance (e.g., lower growth of the "inferior competitor").

When competition is based on the process-based definition, an ecologist may measure the availability, or amount, of a resource that both organisms actually share and measure the amount that each captures when alone and when together. In contrast, when competition is defined as the negative outcome of competitive interactions among organisms, an ecologist may measure the density of population A in the presence of another species B and the density of A when B is not present. A lower density of A when A and B are present together is tallied as a negative sign. Clearly these two uses of the term "competition" are related, but they are not the same. One refers to process and the other to an outcome. The need exists to define one usage as most fundamental to the heart of a theory — process — and the other usage as an operational or conventional measure that may relate in specific ways to either the fundamental concept or to field, laboratory, or model applications — outcome. The potential for confusion in the absence of clear articulation of these differences brings us to the need for theories to include definitions.

D. Definitions

To build a general theory, basic definitions of various terms and objects must already be available or must be supplied (Brandon 1984, Hull 1974, Lewis 1982, Loehle 1987a). Some view the lack of refined terminology of ecology as the main bottleneck in advancing its theoretical and practical strengths (Grimm 1998). Definitions may be verbal or quantitative. As we saw with competition, if the term can refer to more than one idea, it is important to define which manifestation is being used in the current theory or model. Often different terms will have to be defined to distinguish among related conceptual meanings. Thus, definitions can be conventions required to structure the complex conceptual devices of a theory. Definitions clarify, restrict, or prescribe the use of various terms within a particular theory. Furthermore, definitions may label the concepts included in the theory or may be constants required for certain calculations. They may be terms that are unique to the theory at hand and may be derived from other theories.

A complication arises with definitions. Some may be "primitive" terms (Rosenberg 1985, Stegmüller 1976) and not definable by other terms supplied within the theory at hand. Primitive terms can be given meaning by other theories. Donor theories can be broader or more specialized than the theory that adopts the definition. "Fitness" may be considered a primitive term in some versions of the theory of evolution (Rosenberg 1985, Williams 1984). However, the need to use primitive terms may be a consequence of applying the logical positivist view of theory as a deductive system of statements. The statement view emphasized the logically closed nature of theories and so expected definitions to be put in terms that were generated by the theory in question. Under the contemporary concept of theory, such a restriction is not needed. So primitives should present little problem.

Three components must be defined in a theory: (1) objects; (2) interactions, which are the dynamic relationships among the objects (Leary 1985); and (3) the states or static relationships that can exist in the system that is the subject of the theory. Such definitions may arise closely from the assumptions of a theory. One example of a definition is the formula for calculating fitness (W) in a population. In evolutionary theory, the definition of an individual must account for the complexities of clonal organisms, genetic versus phenetic connections, asexual reproduction, and so on. This definition is crucial because it is variation between individuals that is the central driver of evolutionary theory. In landscape ecology, the definition of a patch must account for scale of observation and measurement, as well as the biotic or abiotic parameters that might be used to detect patches. Patches are considered to be areas that are distinct in composition, architecture, or function from other areas, at a given scale of observation. A given kind of patch may be heterogeneous within itself but still differ from other patches at the scale of observation.

BOX 3.4 Pattern and Process: The Phenomena of Ecology

Pattern. Arrangement of entities or events in time or space; confirmable by observation or experiment; exists on a higher level of organization; metaphor for pattern: the "nouns" of an area of study; syntax: pattern is, exists, or occurs

Process. Cause, mechanism, or constraint explaining a pattern; confirmable by experiment; exists on a different level of organization than the target process, to which it is connected by a model; mechanisms are interactions on lower levels of organization, and constraints are causes on higher levels of organization or coarser scales than the target process; cause is a generic term for any influence on a pattern, including both mechanisms and constraints; metaphor for process: the "verbs" of an area of study; syntax: pattern x happens, or is limited, because of process y. Process usually contains invariant and variant components

Phenomenon. Any observable pattern or process

Several of the concepts central to these examples would have to be specified differently depending on the situation, scale, model, or question driving the study. Because ecology deals with contingent systems whose structure and dynamics depend so much on differences in initial conditions, boundary conditions, and the order of events within them, it is important to realize that the definition of a term used successfully in one case may be entirely inappropriate for at least some other cases. As we explained in Chapter 1, definitions are often general so that they can apply to different scales, systems, and processes in ecology. To apply, or specify, these definitions in specific cases, models must be used. This suggests that there is a great deal of care required in stating and using definitions.

To return to our earlier example of competition, once the fundamental concept of competition is identified as the process of joint use of limiting resources, the operational measurements or net effects of competition must be defined and kept distinct to avoid unproductive argument about what competition is and whether it is actually occurring in a particular case. Such clarity and distinctness of definitions is essential to successful comparison across contrasting environments, organisms, and situations (Downing 1991, Heal and Grime 1991). Discriminating pattern and process is one component of clarity in definition (Box 3.4). Thus, competition theory might require, in addition to its central concept of competition enshrined in a definition, more specific definitions of processes that are subsets of competition. For example, (1) interference can be defined as a net effect, (2) allelopathy as the process of interference via chemicals released into the environment, and (3) indirect competition as the result from the net negative effect of a third party that can mediate the competition between two species. In addition, definitions will have to be sensitive to the differences between competition between two organisms and among populations and assemblages of organisms, because competition among some individuals of two species might have a positive impact at the population level. Mechanisms leading to the different outcome on two different realms — individual and population — may involve relative strengths of intra- and interspecific competition or different intensities of competition in heterogeneous habitats. A theory cannot be evaluated, tested, or confirmed if its definitions are not accepted, at least for the purpose of the task at hand.

It should now be clear that a failure to develop good definitions may constitute a major impediment to progress in a discipline. Furthermore, situations where a concept persistently escapes our efforts to define them may indicate serious flaws in the concept itself and thus a need for a deeper analysis of its assumptions and meaning.

IV. Theory and Its Empirical Content

The aspects of theory defined so far are its conceptual foundation. However, theory has an empirical or factual foundation as well. Therefore, any well-developed theory will include some body of accepted facts (Hacking 1983, Lewis 1982, Stegmüller 1976) and the generalizations derived from them.

A. Facts

Facts are confirmable records of phenomena — that is, events, processes, and objects (Mahner and Bunge 1997, Novak and Gowin 1984). Many philosophers recognize that facts have a theoretical component, so concept and fact depend on one another (Hacking 1983). While referring to facts in terms of phenomena, we use the latter only in the sense of states or changes of things independent of the observer and not as perceptions by an observer. That second use is improper and arises from the positivist view of epistemology and is currently viewed as a failed interpretation of what facts are (Mahner 1998). Facts are given meaning by the theory to which they contribute. Facts may depend on some other theory to identify them, and quantify their observation (Amsterdamski 1975). For instance, the "fact" that the sun rises is in reality an interpretation of the observation that the distance between the horizon and the solar disk changes over a particular time period. The observation, highly confirmable and repeatable, is meaningful only in a theoretical context and has radically different meanings in a Copernican versus a pre-Copernican theory of the world.

This example shows, first, that repeatable observations referred to as facts can be temporal or spatial patterns. Thus, in ecology, facts range from the records of distribution and abundance of organisms and fluxes of materials and energy to observations of developmental trends in individuals or assemblages. Second, the example shows that the "meaning" of a fact, such as the sunrise, depends on the theory with which the fact is associated. This is why it is said that facts are "theory laden." Even such simple observations as temperature are highly conceptual. When carefully analyzed, temperature is an observable representing a mean kinetic energy of molecules. Therefore, the apparently simple observable of "temperature" invokes the statistical concept of mean, the physical mechanics concept of kinetics, the fundamental principle of energy, and the chemical concept of molecules. Temperature is, therefore, a sophisticated and complex parameter, and highly dependent on theory.

A simple ecological example of theory-ladenness is the phenomenon of character displacement along environmental or geographic gradients (Brown and Wilson 1956). The basic fact of phenotypic divergence of two taxa in sympatry has differing interpretations in at least two alternative bodies of theory: one is the biogeographic or physical control of body size or other organism features, whereas the other is coevolutionary theory with its implication of evolved difference in niche or reproductive characters driven by competition. In the first theory, displacement might be due to responses of the separate species, whereas in the coevolutionary case, differentiation is a result of interaction between the species. The fact therefore can have different meanings in the two theoretical areas. Differentiating between the two explanations is, one hopes, a matter of test (see Futuyma 1986), but the idea here is that fact and theory intertwine.

Note that the term "fact" can appear in two situations in the discussion of conceptual constructs. The observable phenomena to be explained by a theory or predicted by a theory are also facts by virtue of their potential confirmability. However, for the sake of precision, we differentiate the facts that are already accepted as part of a theory from those that are the targets of new causal explanation, generalization, and testing. These new facts will be distinguished as the observable phenomena within a domain. Note further that observable phenomena — that is observed patterns, states, or processes — once subject to certain tests and generalizations, can become absorbed into a theory as facts. Thus, theory is never devoid of empirical content.

Examples of facts in different theories include the many observations of the ability of species to adapt by changing their genetic structures, the multitude of cases showing the ability of populations to grow under favorable circumstances, the existence of vegetation succession on previously denuded sites, or the energetic openness of populations and organisms. These facts are parts of evolutionary theory, population theory, succession theory, and ecosystem theory, respectively. Examples of observations in need of increased attention include polar to tropical gradients of biodiversity (e.g., Brown 1995), species-area curves (e.g., Rosenzweig and Ziv 1999), or the relationship between diversity and stability (e.g., Valone and Hoffman 2003, Mikkelson 2001). In all cases, more observations of the phenomenon in different situations are needed to better develop or evaluate the theories to which they contribute.

B. Confirmed Generalizations

At some point, the factual basis of a theory will grow so large that it must be condensed. Such condensations are bodies of abstraction that constitute the confirmed generalizations of the theory. Some traditional philosophers do not accept the view that confirmation of generalization is valid (Chapter 2). However, because the utility of generalization in ecology and other sciences is so great (Ayala 1974, Harper 1982, Levins 1966, Longino, 1990, Tilman 1989), it is worth emphasizing that scientifically legitimate modes for dealing with confirmed generalization do exist. It is widely known that Popper (1959) questioned the logic of inductive confirmation, although he later agreed that generalizations may be provisionally confirmed by testing (Popper 1974). Indeed, constructing empirical generalizations always involves implicit tests: accumulation of observations or facts implies the question, "Is the instance now under investigation like the previous ones addressed by the theory?" The conclusion is always subject to revision.

Despite the complaint about confirming generalization, which arises from formal logic, natural scientists have used and continue to use the technique successfully (Brush 1974, Colwell 1984, Gould 1986, Lloyd 1987, 1988, May 1981, Oster 1981). As noted in Chapter 2, the conflict concerning generalization exists because logic deals with the *form* of an argument, whereas empirical science deals with the dialogue between observable phenomena and conceptual constructs — that is, the *content* of arguments. These two modes of argument must be evaluated in entirely different ways. Logical arguments are evaluated by the validity of their form, whereas empirical science is evaluated by content and the relationship between observable phenomena and conceptual constructs.

Recall Lloyd's (1988) analysis of confirmation as a three-pronged attack for evaluating the empirical soundness of theory or certain of its derived components (Chapter 2). The three aspects of confirmation are (1) the degree of fit, (2) the existence of independent support of assumptions, and (3) the variety of kinds of evidence brought to bear. These approaches are all important for determining whether a generalization is confirmed. Vermeij (1987) outlined a similar analysis.

Other ways of looking at confirmation as a valuable and legitimate process have been used by philosophers of science. For instance, the confidence of scientists in confirmation of a generalization increases with the accumulation of observations consistent with that generalization, and confidence in confirmation of a theory increases with the number of different cases explained by the theory (Ruse 1979). The generalization that successional dynamics in plant communities alternate between periods of canopy closure and canopy opening is tentative. Indeed, it is best considered a hypothesis. However, the generalization that succession tends to generate communities of increasing spatial complexity in moist environments where deciduous trees can grow is based on decades of repeatable observations. Such a generalization about classes of facts can be considered confirmed. Confirmation is, of course, provisional.

Generalizations are a broad category with its own taxonomy. Cooper (1998) identified a taxonomical space defined by three dimensions, each of which represents one tendency or type of generalization. These are theoretical, causal, and phenomenological. In other words, the observed patterns can lead to a theoretical distillation of relationships, to formulation of inclusive statements of causes, or to identification of characteristic patterns. This is not to suggest that generalizations emerge in these three pure forms — most often one type informs and interacts with the others. We discuss modes of interaction between various components of theory and how these modes lead to integration of understanding in Chapters 4 and 6.

The confidence in confirmed generalizations also resides, to some extent, in their interaction with other conceptual devices that produce a successful theory (Hacking 1983). For example, if a generalization is used as a key part of a model that successfully and consistently matches observations, the confidence in the contributing generalization would be enhanced. By contrast, the absence of effective dialogue between candidate generalizations and theory impedes progress and generates confusion, as evidenced by the analysis of work on effects of disturbance on coral reefs (Jones and Syms 1998). Confidence also builds because of consistency with other related theories and observable phenomena outside the scope of a theory (Quine and Ullian 1978). The confirmed generalizations in ecology are likely to be expressed as statistical rather than deterministic relationships because of the variability of natural systems and the involvement of multiple causal factors.

The theory of evolution contains a good example of a confirmed generalization. Genetic variation in populations has a fundamental and firm empirical role in the theory because of the overwhelming evidence of its occurrence throughout the living world (Mayr 1982). Examples of confirmed generalizations also abound in ecology. The highly significant relationship between actual evapotranspiration and tree species diversity discussed in Chapter 1 (Currie and Paquin 1987) is a biogeographic example, as are the classical biogeographical rules concerning changes with latitude in body size (Bergmann's rule), length of appendages (Allen's rule), coloration, and so on, of endothermic animals (e.g., Begon et al. 1996). In community ecology, a confirmed generalization is the species-area curve, which states that as sampling area increases in a uniform environment, the number of species encountered increases asymptotically. In succession theory, the concept of communities in equilibrium with local conditions and fine-scale disturbance incorporates a generalization derived from observing a large number of communities long after catastrophic disturbance. In ecosystem ecology, the relationships between primary production and consumption (Cyr and Pace 1993, McNaughton et al. 1989) and between lignin : nitrogen ratio and decomposition rates (Melillo et al. 1982) are also confirmed generalizations.

An important part of accumulating and summarizing cases is adequate classification (Colwell 1984, Gould 1986, Kiester 1980, Loehle 1987b, Mayr 1982, Price 1984, Sagoff 2003, Salmon 1984, Schoener 1986b). A system of classification allows observations to be separated based on similarities and differences. Failing to classify cases effectively or correctly may lead to inappropriate rejection of a generalization. For example, species-area curves will not be asymptotic if the habitats are not classified correctly and, in fact, a highly heterogeneous situation is studied. Similarly, the expected tropical-to-temperate zone gradient in species diversity is not found for all organisms. Some of the exceptions have good theoretical reasons for their status; for example, organisms in certain habitats buffered from environmental uncertainty would be expected to exhibit greater diversity at higher latitudes than those that are exposed to more variable environments. Thus, the fauna of Lake Baikal is richer than the fauna of most other lakes (Kozhov 1963), regardless of latitude. Classifying habitats into different types results in better generalization in this case.

V. Theory and Its Derived Conceptual Content

The components of theory discussed so far have been either straightforward concepts or empirically based content. However, theories also contain important components that are derived from the simpler conceptual components. These derived conceptual components do much of the work of theory and include many of the relationships that exist with the observable world. We therefore move now to two of the derived conceptual components of theory: laws and models. Although laws may be focused on the empirical foundation of theory, they often emphasize or start with the supposition, "What if?" Therefore, laws often express a condition contrary to fact. Laws can also express some degree of causal necessity between two phenomena included in the theory. Models share several of these features with laws but differ because they are complexes of concepts arranged to suggest some outcome of a relationship. These outcomes are intended to describe or explain the world. Specific initial conditions and parameters also play important roles in models. We describe the structure and function of laws and models next.

A. Laws

Laws constitute a special class of generalizations. Laws specify relationships between two or more phenomena, and this relationship may be correlational, or it may be causal. They may be formulated verbally or quantitatively. Quantitative laws may be deterministic or probabilistic (Rensch 1974, Salmon 1984). Laws may have one of two forms. The first form is that of a universal conditional statement: if A, then B (e.g., Brandon 1984, Hull 1974). Note that we use the term "universal" in its sense of applying throughout a specified universe of discourse or domain, not necessarily to the whole literal universe. The universe consists of all instances of A.

Perhaps the best example of a law familiar to ecologists is the law of natural selection (Reed 1981). The law is a universal conditional statement that specifies the conditions under which evolution by natural selection will occur:

1. if members of a biological entity have heritable variability, and
2. if this variability affects their performance relative to an environment, and
3. if they have the capacity for replication in excess of the capacity of the environment to support them,
4. then progeny of those members that vary in closer conformity with that environment will accumulate in subsequent generations.

Ghiselin (1969), Rosensweig (1974), Gould (1977), and Mayr (1991) provide equivalent statements. The law of natural selection is literally universal in one sense because it applies to any collection of entities anywhere in the universe, if their attributes are heritable and variable, and they are subject to an external limit to their expansion (Ghiselin 1969, Reed 1981, Williams 1970). It does not, however, apply to all entities in the physical universe, but only to those that meet its assumptions. We call this statement a law because of its universal conditional form, not because of its high degree of confirmation. Indeed, the law of natural selection has no empirical content that could be confirmed (Brandon 1990). Its empirical relevance is provided by filling in the blanks, so to speak — that is, an environment must be specified and the biological entity of interest must be specified. Then the degree of match between performance of the entity and the environment, and the validity of environmental limitation, can be empirically tested (Brandon 1990). Testability and confirmation reside in the application of the law to specific instances.

The theory of succession contains a similar central law. This law can apply to any community, although it is framed in terms of plants. It adopts the same form as the law of natural selection (Pickett et al. 1987):

> If an open site becomes available, and if species become available differentially at that site or species perform differentially at that site, then vegetation structure or composition will change through time.

Both laws presented here are multiple conditionals. Multiple conditionality is appropriate in ecological laws because ecological relationships are recognized to be highly contingent. Contingency implies that several to many factors govern the outcome of an ecological process, and the order in which factors act can determine the outcome. A different mix or different order of the same factors can lead to a different outcome. In cases where contingency depends on multiple factors, specific models or subtheories must be used to complete the description, prediction, or explanation (see "Translation Modes"). Like the law of natural selection, the universal conditional law of vegetation or community dynamics derives its empirical content from application to specific sites, environments, and collections of species.

Natural laws can take a second form, that of an identity. Some identities can hide tautologies. A tautology is a statement that is true by virtue of its form: P or not P. This little bit of logical jargon means that a phenomenon has property P or it does not have that property. This statement is logically unassailable, but it tells a scientist nothing empirically useful about the phenomenon of interest. Therefore, tautologies are obviously not of interest in empirical science because they have no relevant empirical content. However, some truly useful identities exist. The value of laws having the form of identities is in linking things or processes that might not appear to be related, and doing so in some functionally significant way that can be empirically evaluated. The process laws of physics, for example, have the form of identities and equate a future or distant state with a modified current or local state (Hull 1974). Laws can define states, specify outcomes, or specify causes.

Figure 3.2 summarizes various aspects of laws that have conditional form and reflects differences of opinion on what a law is relative to an observation, how we find laws, and what we can

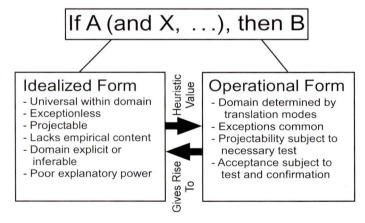

Figure 3.2   Two kinds of conditional laws.

expect of them. There seems to be both a strong and a weak version of the concept of law. These versions are not mutually exclusive but neither are they entirely compatible. Both are useful, but because of disagreements as to which is correct and appropriate, ecologists may have difficulty deciding on what their laws are. The stronger, idealized version is a more effective tool for theory advancement. However, the weaker version is closer to practice in ecology. The two versions may interact in the process of idealization and translation. The two forms differ in the kind of evidence they require. The idealized form is likely to gain acceptance when its assumptions are undisputed and logic is flawless. In contrast, the employable form is likely to be accepted when the quantity of evidence is overwhelming and exceptions well understood. The search for laws in ecology will benefit from clarity as to which version of law is postulated. The examples of candidate laws that we provide (Box 3.5) represent the idealized form.

Lange (2005) further refined the search for ecological laws by supplying criteria delineating laws from other generalizations. Specifically, he pointed out that a proposition is a law when it is *stable* — that is, when it stays true within the factual and conceptual scope of ecology and does so in a necessary manner. That is, the relationship is not an accident of history.

Because of the philosophical debate about the precise definition and understanding of what laws are (e.g., Cooper 1998, Sober 1993), ecologists may remain unable to formulate ecological laws with logical firmness matching the well-known laws of classical physics (Box 3.6). Notwithstanding, the role of general, conditional statements in developing theory and generating explanation is central to a successful intellectual effort in ecology.

We turn now to examples of laws that have the form of an identity. One ecological example of an identity as law is the thermal energy budget, which describes the fluxes or contents of thermal energy that an ecological entity can experience:

$$(S_i + S_o) + (L_i + L_o) + (H_i + H_o) + (E_i + E_o) + M = 0,$$

where S is shortwave radiation, L is longwave radiation, H is sensible heat transfer, E is latent heat transfer, and M is energy stored in the system (Lowry 1967). The subscripts i and o refer to input and output of energy from the entity or system, respectively. The law basically states that the thermal energy inputs and outputs of a system will ideally balance one another. Rearrangement of the equation permits different components of the energy budget to be varied either experimentally or abstractly to explore the behavior of organisms that are out of thermal equilibrium. The law can also be applied to other ecological systems when the equation is parameterized differently to take account of their three-dimensional architecture.

An additional important example of an identity law is the Hardy-Weinberg principle (H-W principle). This law is a key aspect of the theory of evolution. It states that, in the absence of mutation, genetic drift, selection, migration, and selective mating, the frequencies of genotypes *AA, Aa*, and *aa* will be distributed as p^2, $2pq$, and q^2, respectively, and will sum to unity in a large, sexually reproducing, diploid population. The principle indicates the specific conditions under which no evolutionary change, at the genetic level, will occur.

The H-W principle functions as a "zero force" law in evolutionary theory, analogous to Newton's first law of motion (Sober 1984). Zero force laws indicate the conditions under which no change or alteration of an established trajectory will occur. If the conditions for the H-W principle are met, there will be no net change in the genetics of the population. If the assumptions of the H-W principle are *not* met, evolution will occur (e.g., Wilson and Bossert 1971). Thus, this law suggests the mechanistic details necessary to fill out the theory of evolution, in addition to those embodied in the law of natural selection (Fig. 3.3).

Zero force laws are important components of many theories that deal with dynamic phenomena, as many ecological theories in fact do. They provide a useful calculation device or an ideal reference point or trajectory against which to compare and simplify the complexity of the natural

BOX 3.5 General Statements That Meet the Conditions for Being Laws in Ecology

We submit this list of candidate laws as a tentative answer to the question posed by some ecologists and philosophers (Brandon 1984, Lawton 1999, Murray 2000, Quenette and Gerard 1993, Turchin 2001): "Does ecology have laws?" Our hope is that it will stimulate further discussion and analysis. Most of the laws supplied here can be classified as "schematic" (Brandon 1990), implying that they have a high organizing power but need further specification before being employed in practice. In this they are similar to some laws of physics.

- A population with constant age-specific rates of survival and initial size of cohorts maintains a steady state (Murray 2000); but see Turchin (2001) for serious criticism.
- In the absence of changes in age-specific birth and death rates, a population will eventually establish a stable age distribution (Murray 2000).
- If the environment experienced by each individual of a population remains constant, the population will change exponentially (Turchin 2001).
- If a system is a pure consumer-resource system (in which per capita rates of change of both resource and consumer do not depend on their own density), it will inevitably exhibit unstable oscillations (Turchin 2001).
- If two species occupy a homogeneous environment, and if those species have congruent niches, then they cannot coexist at equilibrium (see Chapter 5 for an analysis).
- If an open site becomes available, and if species become available differentially at that site or species perform differentially at that site, then community composition will change through time (cf. Pickett et al. 1987). The process of change will stop when differential performance no longer takes place.
- If the collection of organisms constitutes a trophic level L and these organisms feed on another collection of organisms at trophic level L-1, the production P at a trophic level L is related to the production at the trophic level L-1 according to inequality $P_{L-1} \gg P_L$.
- If two ecological entities (individuals, kin groups, populations, species, ecosystems) are exposed to the same set of conditions, they will respond differently.
- In the absence of evolution, any nonisolated habitat will asymptotically approach an equilibrium number of species, with the equilibrium value being a positive function of the colonization rate and the habitat's ability to support the species that arrive.
- If an organism of body weight W has a total metabolism of M, and the organisms are of the same kind, then another organism a times greater than W will have the total metabolism approximately proportional to $M = a*W^c$, where c < 1 (modified after Colyvan and Ginzburg 2003).
- Every ecological system in which repair occurs at a slower rate than damage will be replaced by a simpler system (in terms of number of interaction types, body sizes, species diversity, etc.).

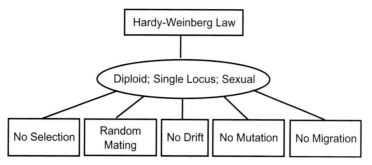

Figure 3.3   A schematic diagram of the Hardy-Weinberg law. As originally conceived, the equilibrium, or zero force law, applies to sexually reproducing, diploid organisms, and is illustrated by a single locus with two alleles. Although later expansions of the law accounted for multiple alleles and sex linkage, the simple form, $1 = p^2 \cdot 2pq \cdot q^2$, suffices to illustrate the nature of a zero force law. For the Hardy-Weinberg law to hold, five assumptions must be met: an absence of selection, genetic drift, mutation, migration, and the existence of random mating in the population. Under other conditions, evolution in the form of altered gene frequencies from generation to generation will occur.

BOX 3.6 For Discussion: Debate on Laws in Ecology

A number of objections or arguments for ecological laws have been presented by both ecologists and philosophers of science. The debate does not appear resolved but it may be important to the determination and pursuit of ecological questions and ultimately to the success of ecology.

Let us begin with an assertion that the existence of emergent properties precludes obtaining high-level ecological laws by just deducing them from those of other sciences (Bunge 2003). Hence, ecological laws must be deduced at the appropriate levels of ecological organization, involve ecologically based assumptions, and use logic or known ecological relationships to organize the assumptions into a prospective law. For example, the mechanisms by which plants or animals compete are entirely different from those by which atoms compete. While plants process energy and matter consistent with the laws of physics, it does not mean that the dynamics of competition are directly deducible from physical knowledge (Marone and Del Solar, in press, Murray 2000). A simple example illustrates this point. When lions and hyenas compete for a carcass, the outcome determines where the energy is going to be allocated and in what quantities. However, neither the energy of the carcass, the energy of hyenas, nor that of lions appears to be informative of the outcome of competition. Thus, competition is a phenomenon that appears at the level of a system that consists of at least two ecological entities (i.e., populations) and cannot be deduced from nor rigorously linked to the principles of physics.

Having asserted that laws of ecology are not those of physics, the question arises of why ecologists seem to have difficulty with formulating and agreeing to what the laws are. The most common putative cause invoked by ecologists, as well as some philosophers of science (Shrader-Frechette 2001), is the contingency of ecological phenomena (Knapp et al. 2004, Lawton 1999, Marone and Del Solar, in press). Sagoff (e.g., 1997) and Shrader-Frechette and McCoy (1993) have repeatedly condemned ecology's pretensions to nomotheticity, or law-likeness, of some of its regularities. Ecology is too complex, they have said, to be fruitfully characterized in terms of general laws (Mikkelson 2003). However, Simberloff (2004) cautioned after Windelband (1894) that ecologists should not confuse the distinction between general laws about the structure and workings of nature with idiographic knowledge, which depicts singular events and focuses on unique aspects of particular phenomena — the sources of complexity.

Complexity is then seen to derive from contingency, and contingency itself is seen as a condition where any particular outcome depends on a number of contributory causes that can act in a fairly unpredictable mix. Irrespective of the level of organization, contingency is believed to be fundamentally different from that of physical phenomena and hence an insurmountable obstacle to finding meaningful laws in ecology. However, this claim seems to be taken on faith and, when confronted with reality, borders on absurdity. The great majority of physical phenomena is as contingent as biological ones, and some are entirely indeterminate. Whether one considers a flight trajectory of a falling object, the distribution of rocks on the slope of a mountain, movement of air particles or air masses, locations of lightning strikes, arrangement of matter in the universe, spread of fire, shape of a snowflake, or hundreds of other physical phenomena, contingency is pervasive. Rather, it is a matter of abstraction and idealization (see Chapter 2), not of subject, that differentiates the constructs of physics as compared to constructs of ecology, at the moment at least. This means that laws of physics are usually formulated as if no other forces or modifying factors existed. Laws of physics rely on stripping the contingency to expose an ideal relationship, a relationship that is difficult to observe in nature. Consider the simplest and best known law of physics, that of the first law of mechanics. The first law deals with forces and changes in velocity. For just a moment, let us imagine that you can apply only one force to an object — that is, you could choose to push the object to the right or you could choose to push it to the left, but not to the left and right at the same time, and so on.

Under these conditions, the first law says that if an object is not pushed or pulled, its velocity will naturally remain constant. This means that if an object is moving along, untouched by a force of any kind, it will continue to move along in a perfectly straight line at a constant speed. The operative phrase is *under these conditions* (i.e., of one force only). When more forces apply, as always is the case, the first law of mechanics will fail in its predictions. However, physicists are happy with this law. Should not ecologists be able to construct similar laws and be happy with them?

In spite of the reasons to the contrary, ecologists worry about the contingency and its negative effect on their ability to synthesize multiple streams of observations. One attempt to come to rescue (Knapp et al. 2004) starts by conceding to critics that laws may not be attainable because of fuzziness of relationships among ecological entities and phenomena. Knapp et al. (2004) postulated a scaled-back program for ecology — a program that will focus on finding rules. They seem to subscribe to the collective arguments and logic of Lawton (1999), Simberloff (2004), or Berryman (2003) and find comfort in the fact that most ecologist agree to the existence of rules. They defined rules as generalizations or statements that predict the occurrence of a particular ecological phenomenon if certain conditions are met. However, their definition of rule is not much different from the definition of law. So what is the problem? The quality of prediction or the quality of the formulation?

One might argue that the search for rules might be good because it could inspire finding laws. One might also argue that the search for rules might hinder progress by emphasizing empirical over conceptual work. Newton's first law of mechanics illustrates these two possibilities. The law says that an object pushed should move at a constant speed along a straight line. Although no object obeys this law in the natural world, the fact that most objects tend to move in one predominant direction and continue to do so for a while might lead a speculative observer, as it did, to the formulation of the law. Hence, empirical observations summarized as a rule that movement occurs, at least initially, in a direction not much different from a straight line and continues for a while at least after the force stopped suggested a general conditional rule that would only work in an idealized setting — a law in short. However, these same observations might lead one to a proliferation of rules such that, for example, (1) fluffy objects show greater trajectory variance than do dense objects (like bullets), (2) objects whose initial speed is less that the first cosmic speed tend to fall to Earth, (3) objects whose initial speed is greater than the first cosmic speed tend to orbit Earth, (4) objects in water float if their density is less than that of water or sink to the bottom if their density is greater than that of water, and so on. All of these rules, and many others, are useful, true, and would represent progress in recording and understanding natural phenomena. The question ecology faces is not which of the strategies to pursue but what is the most promising mix of strategies, all of which should be pursued. The claim that laws cannot be found and cannot be useful in ecology (see Mikkelson 2003) has little basis in logic and the practice of science in general.

world. The example of the H-W law offers a chance to say what laws do not need to do. According to Colyvan and Ginzburg (2003), laws need not distinguish between cause and effect, need not have explanatory power, nor need to be predictive in any specific way. These observations can aid in resolving some current debates in ecology about the plausibility of formulating general laws (e.g., Lawton 1999, Murray 2001, Turchin 2002).

The difference between the different kinds of laws points out a difference in the kinds of systems that scientists can study. In closed systems, such as those in classical physics, process laws allowing the inference of a future state from the current state of the system are often encountered (Hull 1974). In contrast, in open systems, like those often encountered in ecology, laws may specify the inputs and relationships needed to describe, predict, and explain specific ecological phenomena under various conditions. Thus, ecological laws may be multiple conditionals (e.g., if A and B or C, then D) and can account for history as well as a current state by projecting into the future or over space. For example, we can formulate a tentative or preliminary law relating the effects of the environment on plant resistance to insects

1. if the external abiotic environment changes, and
2. if a plant is sensitive to these changes, and if the plant response alters tissue biochemistry, and
3. if a specific performance parameter of an insect herbivore is sensitive to these biochemical changes, and
4. if the insect life history characteristics enable performance changes to translate into changes in population abundance, and
5. if there are no other constraining forces on herbivore population dynamics,
6. then herbivore abundance will change.

Note that this law is relatively more complex than the law of natural selection or the law of succession. As laws are established to deal with more specific interactions, they may have to specify more phenomena. In the cases of succession and evolution, application to specific cases is accomplished through adding models to address those situations, not by making the basic law more complex. Indeed, ecologists usually deal with specific domains through models rather than through general laws. This points out how these two components of theory can differ from one another.

Both multiple conditionality and probability, as in the statement of the plant-herbivore defense law, are likely to be widely encountered in ecological laws. Note that the laws of classical physics have an unstated assumption that the objects of interest can be considered well delineated and reducible to idealized mass points between which material interactions are directional, instantaneous, with one-to-one mapping of cause and effect. Laws having the same formal structure as those of classical physics may be problematical when applied to concrete biological phenomena because of the obvious mismatch. Ecological phenomena are often not spatially discrete, not idealizable because of history and contingency, subject to bidirectional causes or feedbacks, and have complex, multiple causality (Haila 1986). These are some of the reasons that multiple conditional laws are likely to appear in biology.

While we have pointed out some of the distinctions ecologists and philosophers make between laws of physics and prospective laws of ecology, we must caution the reader that these distinctions must be treated as quantitative rather than qualitative. Physical phenomena as simple as objects falling to the ground are also affected by many factors. A thought experiment illustrates the difficulty. If a committed experimentalist threw different objects out the window of 10-story building, she or he would find that almost none of the objects behave in conformity with the laws of gravity. Whether it is paper, feathers, lead balls, or a boomerang, each object would fall at different speed and acceleration, and would follow a different trajectory or not fall immediately at all. This is because each object is affected by other forces such as wind, its own aerodynamic lift, and gravity in a unique combination. Contingency rules. Clearly, a law of physics shows weaknesses that are commonly believed to be typical of ecology. Perhaps the differences between physics and ecology are overrated.

One difference in laws from different sciences may require a separate brief discussion as it is sometimes raised as an explanation for the difficulties ecology encounters in the formulation of laws. To ecologists, the simplicity of physical laws stems from the observation that they describe either single items or statistical behaviors of many identical items (gas molecule versus gas as a substance). This simplicity seems to contrast the uniqueness of individuals, populations, or species. However, there is no logical principle that would prevent science from formulating laws that capture behaviors of systems composed of a multitude of different items. Such a challenge may turn out to be a more difficult one, but this difficulty may only be daunting as long as ecologists try to follow the physics model and give up on an ecology model of science. The latter might need to focus on regularities and laws that rule over the interacting collections of unique items or entities.

An important feature of laws is their generality and force. Some philosophers see no laws in biology (Bernier 1983), whereas others assert that biological laws exist (Colyvan and Ginzburg 2003, Ruse 1979, 1988, Sober 1993). Cooper (2003) discussed the nature of laws in ecology and suggested adding an important criterion to the evaluation of laws. Laws must be evaluated by their degree of lawfulness or nomological power. Such power arises from the scope, confidence, and the role of law as component of broader theoretical framework. In other words, how a law fits into a theory is a criterion for its lawfulness. It is impossible to evaluate the utility or appropriateness of a law in isolation from a theory.

An evident danger is that every relationship in biology could end up a law. There are two ways out of this dilemma. One is to restrict the construction of laws to highly abstract and, hence, likely generalizable relationships. Under this rule, commonly or directly observed phenomena would not be the subjects of laws. For example, in another field of science, gravity was not observed in order to generate the Newtonian laws; acceleration was a highly abstract parameter, derived from simpler, readily observed rates. So even in physics, not all important principles exist as laws. The other solution is to codify the universal relationships in ecology but rank them in importance or scope by using their position in the hierarchical frameworks of theories as a guide. The implications of this second suggestion must wait until the hierarchical structure of theory is explained in a later section of this chapter.

One final relationship exists between "theory" and "law." In the Baconian view of theory as an inductively verifiable collection of statements (Carnap 1966, O'Hear 1990), theory development was summarized by the following sequence: hypothesis, theory, law. Because of the abandonment of the statement view of theory by philosophers (Hacking 1983, Rosenberg 1985, Stegmüller 1976), we cannot consider a confirmed "law" to be a "theory" in some more mature manifestation (Amsterdamski 1975). Still, this inductive chain appears in some elementary accounts of science and its methods. It should be discarded. In fact, more recent analyses (e.g., Mahner and Bunge 1997) impose a condition for a statement that it be a part of a theory before it can qualify as law.

B. Models

This section discusses models and their role and status within theory. Models, as components of theory, are conceptual constructs that represent and simplify reality by showing the relationships between the objects of a theory, the causal interactions, and the states of the system (Nagel 1961, Starfield and Bleloch 1986, Suppe 1977a). Models may be verbal, quantitative, graphical, or physical. Note that models differ from both simple concepts and compound concepts because not only are models compounded from other concepts, but they also involve a high degree of abstraction or idealization, have internal logical structure, and generate empirical implications. In a sense, models are the explicit working out of the notions, assumptions, or concepts mentioned earlier, or they may be derived from confirmed generalizations and laws. Models may be the principal mode for generating expectations from some theories (Nagel 1961). Models, therefore, extract the essence of a situation, a relationship, or an entire theory (Kareiva 1989, Loehle 1983, May 1981). However, even though models may get at the essence of a theory, it is important to realize that models are not the entirety of theory (Lewis 1982, Pielou 1981, Stegmüller 1976). Much of the informal parlance in ecology violates this distinction and speaks of models and theory as equivalent. In the contemporary view, theories can be considered to be families of models (Lloyd 1988, Thompson 1989). However, we have suggested earlier that theory has an empirical content as well, which makes it substantially more than just a family of abstract models.

Not all models will include all the aspects just listed; those that omit causal interactions, certain relationships, or certain states are less general or complete. For example, incomplete models may be perfectly capable of describing and forecasting dynamics, but they may not be useful for generating predictions in novel situations. Likewise, a model omitting an object or component of a system may function adequately under some conditions but not under those in which the neglected parameter becomes effective. For example, a model assuming unlimited dispersal as the rule fails to explain the dynamics of dispersal-limited situations in marine intertidal communities (Roughgarden 1989). Again, classification of systems or cases and comparison with the assumptions of the models is a critical step.

Several types of models, all fitting the basic definition just given, can be recognized: (1) static, (2) system or functional, (3) analytical, and (4) simulation (Levin 1981, Levins 1966, Loehle 1983, Pielou 1981, Starfield and Bleloch 1986). These different kinds of models have different structures and uses.

Static models are essentially pictures of system structure. A profile diagram of forest canopy layering or a map of a landscape is a static model (Fig. 3.4). Such models do not illustrate any

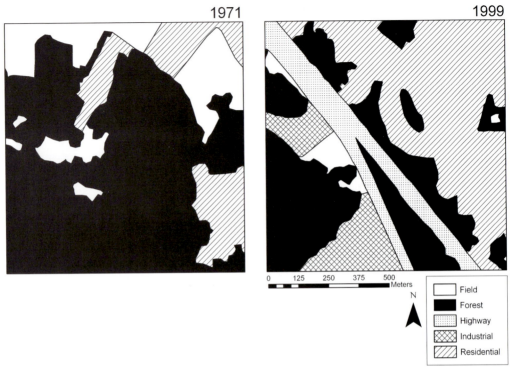

Figure 3.4   A sequence of two landscape maps, or models, of a portion of metropolitan Baltimore, Maryland, in the United States, from 1971 and 1999. Although each map is a static two-dimensional represention of a landscape, combining the two shows landscape change. In particular, the reduction and fragmentation of forest cover at the expense of suburban residential expansion, highway, and industrial covers occurs over the 28-year period. Data from the Baltimore Ecosystem Study, Long-Term Ecological Research Program. Courtesy of M. L. Cadenasso.

changes the system might undergo, but they do represent the physical relationships of the parts of the system. A system model emphasizes the flows of materials or pathways of functional relationships among the various components of a system, for example, a community food web. Thus, in system models, the effects of certain components of the system on other components are taken into account explicitly in the structure of the model. Examples of system models include nutrient flow diagrams (Fig. 3.5).

Analytical models are those constructed of mathematical formulas or operations, in which the conclusions arise from the equations employed. The well-known logistic population growth models are analytical (Box 3.7). The behavior of the system is understood by analyzing the behavior of the equations constructed to represent it. Simulation models are rule-based constructions to generate new states of systems triggered by the passage of time or the occurrence of some event. They are intended to mimic closely the behavior of a system. Since system behavior may be mimicked by many different model structures and assumptions, the structure of a simulation model does not have to represent the causes of change in the real system. Consequently, frequent criticisms of models by ecologists for their lack of realism are misdirected. Models, as a component of theory, have their job defined by the theory. Sometimes, a theory underpinning the model may call for a realistic model, but this is a special rather than a general case.

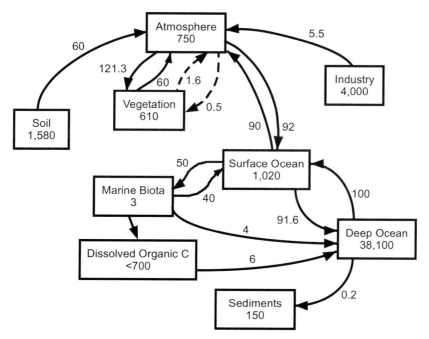

Figure 3.5   A model of the global carbon cycle. The major forms and pathways of carbon in the global ecosystem are shown. Pools are shown in gigatons (Gt) and fluxes in Gt per year. Redrawn after NASA (http://earthobservatory.nasa.gov/Library/CarbonCycle).

Here we pause to present an example of how models link with the components of theories discussed so far and which emphasizes that a single term in ecology can refer to different components of theory. The example suggests flexibility in ecological terms, along with the need for cautious use of those terms and understanding their relationship to theory. The example we will use is how the ecosystem concept is translated into models (Pickett and Cadenasso 2002). First, the core definition of the term brings an idea to bear from outside of ecology. The concept of system, an interrelated set of enties, is the founding notion for the theory. Tansley (1935) took this physical idea as a metaphor and generated a specific definition for ecology. The definition grew out of dissatisfaction with the theory in vegetation ecology based on the assumption that the study objects were organism-like aggregations of plants. Tansley (1935) defined an ecosystem as the interaction between a biotic complex and an abiotic or physical complex in a specified area. Thus, the plants, microbes, and any animals that might be present interact with the soil,

---

BOX 3.7   The logistic equation as an example of an analytic law

$$dN/dt = rN(K - N)/K$$

where r is the per capita rate of increase, t represents time, N, the number in the population, and K, the carrying capacity. The structure of the equation determines the form of the relationship among the components of the system.

water, and atmosphere in a circumscribed area. The idea can be applied at any spatial scale and for any kind of habitat, including both aquatic and terrestrial. But in order to apply such a broad and abstract concept as embodied in the definition, a part of the material world has to be specified. Therefore, the first job is to define the boundaries of the ecosystem to be studied or represented. This is the first step in specifying the domain to be considered. Further specification of domain will delimit exactly what organisms are to be involved and how they will be represented. Will the organisms be quantified as species or as functional groups? Will the abiotic sphere be aquatic, terrestrial, or some combination? The statement of domain will also determine what time span will be examined and how often the system will be examined. This last specification is the statement of temporal grain. The specification will also have to indicate what interactions exist among the organisms and between the organisms and physical environment. Will nutrient flow be considered, or will energetics be the topic? Other interactions are possible foci as well. Once the interactions are chosen, the currency that supports the interaction must be stated. Whether the boundaries of the system are relatively open or are closed with respect to the currency can be either assumed, or it can be a matter of empirical study — that is, openness may be assumed or hypothesized. Assumptions like that of openness must be validated or tested. The rate of exchange or transformation of the currency may be controlled by a variety of factors. Which one(s) of these will be included in the model is a further strategic decision the researcher must make. Hence, specific constraints and drivers that affect the currency become parts of the model, and the mechanism(s) by which they work can be specified as well. At this point, the model may be sufficiently precise to allow the behavior of the system to be simulated, measured, or altered.

This example is but one way that the general, core definition of ecosystem may be converted or specified as a model. As we have seen, along the way, assumptions will be stated, a domain established, concepts to be used identified, parts of the system determined, and interactions suggested. By incorporating known facts about how the interactions can be controlled with observations of the levels or changes in those controlling factors in the system, the central mechanisms behind the dynamics of the system are laid out.

Like the concept of ecosystem, the general concepts in ecology — such as succession, competition, invasion, among others — can be considered scale free and open to a variety of specific mechanisms and trajectories that may appear in specific systems. The biggest concepts in ecology are likely all to be idealizations of this sort, which are given life in real, simulated, or experimental situations through the specification of models. This is the essence of ecological modeling, regardless of the exact forms it may take. The linkage between abstract, general, or idealized concepts and their application in a model is a large step for which many things must be specified. Definition and specification therefore appear to be two of the big jobs of theory in ecology. In informal terms, therefore, theories may be said to deal with "meaning and model" in ecological systems.

C. Theorems

Theorems are derived constructs, deduced from the axiomatic structure of certain theories (Rosenberg 1985) or models. They are usually designed to advance the logical structure of a theory. The status of theorems should be viewed in a context: a statement that is derived in one theory may be a foundational (basic) law in another (Mahner and Bunge 1997). Theorems in the traditional sense are most likely found, along with axioms, in mathematics, geometry, and logic rather than in the empirical natural sciences although there exist tentative efforts to use them \in ecology (Turchin 2001). The ideas of axiom and theorem have traditionally been closely associated with the discredited statement view of theory, the concept of theories as systems of statements (Carnap 1966, Nagel 1961, Rappoport 1978). This conception originated in pure mathematics, for which it is, of course, appropriate (Stegmüller 1976). In the natural sciences, for which the statement view of theories is less appropriate (Grene 1985, Hacking 1983, Salmon 1984, Shapere 1974), theorems in the formal sense may be absent from most theories and may be replaced by other forms of inference (Brush 1974, Gould 1986, May 1981, Suppe 1977a).

In ecology, strict theorems are embedded in mathematical models or in the most general and abstract of theories. Theorems would rarely appear elsewhere in ecology (but see Ford 2000 for a more extensive use of theorems and axioms). Although theorems will be of less importance in natural science than in logic or mathematics, since such strictly logical deductive statements may be important *links* within theories, we retain the term to specify such internal links. When a derived statement has empirical implications or content, it functions as a generalization or a hypothesis.

D. Translation Modes

Translation modes are required to relate the abstractions made by laws, generalizations, and conceptual or quantitative models to the field or to experimental systems relevant to the theory (Levins 1966). Transferring general or abstract theory directly and literally to a field or laboratory case can be problematical (Levin 1981, Oster 1981). How will the concepts contained in the theory be measured? How will change be detected? Translation modes allow ecologists to deal with the contingency of their subject matter while being guided by the clean, clearly derived components of theory. The mode of translation may be a more specific model or a subtheory of the more abstract one (Grant and Price 1981, Levins 1966, Lewontin 1974, Suppe 1977a). In the case of a more specific subtheory contributing to translation, there is still more work of translation to do. Translation is therefore a layered, hierarchical process.

Incidentally, the term "translation" is a holdover from the days of the statement view of theory. If, as the positivists thought, theory was a series of statements that allowed one to translate a generality to the material world, of course the tool would be another statement or series of statements that simply translated from the abstract to the concrete. With the demise of the statement view of theory, we can see how this simple idea of translation modes being subtitles in the abstract movie of ecology to the life outside the theater would fail. But if we can transfer some of the sense of translation from its original context of theory as statements to the context of theory as a conceptual system, the basic idea is still valuable. Abstraction and idealization have to be translated into concrete models, operational terms, feasible measurements, plausible methods, and interpretations of experimental outcomes.

Successful use of theory for integration in ecology may require increased attention to translation modes. This is because one way to match disparate theories is to seek congruencies between them. Congruence in level of detail is one of the principal ways in which theories to be joined should match (Cadwallader 1988). Alternatively, theories can be combined if one adds the more general context of one to the constraint suggested by another. Likewise, combining theories may add the detail and mechanism of one to the generality and idealizations of a different theory. In both cases, translations between the different levels of detail or abstraction of the two theories must be made.

Translation modes have other jobs in theory as well. Translation modes can relate to the testing of theory. Without construction and use of proper translation modes, the tests of model output or hypotheses can be misleading. Translation modes may specify conditions of a test or interpretation or may identify additional factors that must be added to a model to confront a particular situation. Translation emphasizes that theories have core abstractions, which more and more closely match the observable world as additional translation modes are employed. Adding more detail appropriate to the observable world and interpreting the idealizations in terms that better match contingent, messy reality are modes of translation. The problem of translation modes appears, in a way, as the problem of moving across levels of abstraction, ranging from the general, most abstract theory to specification of the local, the empirical, and the experimental.

Translation modes are quite diverse and, therefore, difficult to characterize in the abstract. A specific example appears in the application of the generalized alternative models of succession proposed by Connell and Slatyer (1977). The conceptual system consisted of three "models": facilitation, in which early arriving species promoted the growth and reproduction of later species; tolerance, in which the species followed one another in order of tolerance of the environmental resources and stresses; and inhibition, where early arriving species in fact inhibited the later successional species. These alternatives were a major conceptual advance at the time, but they proved difficult to test in the field (Armesto and Pickett 1985, Hils and Vankat 1982, Pickett et al. 1987, Walker and Chapin 1987). The conceptual system was important because it prevented ecologists from thinking that there was only one way that succession occurred, and it prompted a flurry of experimental work to discriminate among the alternative models. It turns out that the three "models" of Connell and Slatyer (1977) are descriptions of three different net effects that can occur as a result of multiple interactions in succession. However, the specific concepts and mechanisms necessary to use the generalized net effects of facilitation, tolerance, and inhibition (Connell et al. 1987) in the field were not specified in the original theory. For example, Armesto and Pickett (1985), in an experiment to apply the Connell and Slatyer models in postagricultural old fields, had to specify community structure, resource release, and timing of disturbance relative to life histories of species in the system to interpret the mechanisms and outcomes of interactions. This exercise led to the articulation of ways to translate from the generalizations of Connell and Slatyer (1977) to the specifics of experimental tests. The direct application of the logical alternatives of Connell and Slatyer to the field was impossible. An intermediate conceptual model incorporating additional specific factors was needed. Note that the hypotheses in this case were suggested by the high-level general concepts of Connell and Slatyer (1977), but the modes of translation that allowed the legitimate test of the hypotheses were lower level tools of the theory of succession.

Another example of a translation mode comes from insights into plant defense theory. Insect damage on one leaf of a plant can cause other leaves to become more resistant to subsequent insect herbivory. However, chemically induced resistance against insect herbivores following damage often shows inconsistent patterns among leaves. Once it is recognized that some leaves are highly interconnected via specific vascular traces, whereas other leaves on the same plant are weakly interconnected, the variation in induced resistance can be explained. Specific knowledge of vascular architecture is a necessary translation mode for comparing induced resistance responses among plants (Jones et al. 1993).

A whole array of translation procedures may be used in any given case. A test of one of the abstract tenets of hierarchy theory illustrates the structural complexity that can emerge in testing theories. The tenet is that entities belonging to lower hierarchical levels change or operate at higher rates than those belonging to the levels above them. Waltho and Kolasa (1994) identified a concrete system in which to conduct the test — an archipelago of patch coral reefs. In the second step, they abstracted a hierarchical structure of the community inhabiting the reefs. Their abstraction was to restrict the hierarchical structure of the whole ecological system to that of its fish components. The hierarchical structure of the fish community was then assessed using an *auxiliary model* of habitat structure (Kolasa 1989). At that point, these investigators adopted two measures of observables that were to represent some of the many possible attributes of the entities. These measures were abundance of individual species and their ecological range. Both measures were further *operationalized*. Abundance was adjusted by the time individual species spent on separate reef patches. This adjustment eliminated double counts of mobile fishes. Ecological range was measured by the number of different patches and their cumulative area. Finally, the variation in these two parameters was *assumed* to be a correlate of change in the entities present at various hierarchical levels (and fish were viewed as facets of the entities constituting the hierarchial organization of the system). Each of these steps introduced biases and additional assumptions, which, ideally, should all be evaluated.

As we have seen in the reef application, critical to the translation of theory to field or laboratory are concerns of operationalism. How can the ideas and tools of a theory be translated to the concrete or specific situations? In addition to clear questions needed to guide such connections, it is also necessary to know when variables are relevant and how they are to be measured (Keddy 1987, Lewontin 1974, Peters 1980). Such concerns apply to many of the components of theory, but they resonate most clearly with concepts, models, and translation modes. At the most practical level, methods often represent translation modes, fully or partially. In population ecology, for example, methods for collecting age structure data and frequencies are a necessary link extended from the theory of population dynamics to field observations. Counting individuals of mobile populations, of secretive organisms, of nocturnal animals, or of especially long-lived populations often involves assumptions that are required because not all organisms can be censused. In such situations, methods also become components of theory (e.g., Ratner 1990).

E. Hypotheses

The various conceptual constructs of a theory — especially the assumptions, generalizations, laws, and models — will suggest, or can be cast as, hypotheses. The principal motivation for building hypotheses is the need to build good theories (Brandon 1984). Hypotheses are explicitly testable statements. Note, however, that the most useful and compelling hypotheses are those that are clearly connected to a theoretical context (e.g., Mayr 1996). If a person proposes a hypothesis that is entirely disconnected from established or emerging theory, it may not be terribly useful to act on that hypothesis (Quine and Ullian 1978). Some theories may suggest hypotheses that are so grand that they constitute a test of the whole theory (Stegmüller 1976). Such a situation is most likely to exist (1) early in the development of a theory, (2) for theories that suggest a radical departure, or (3) for theories with a narrow focus. More often, especially in the case of highly synthetic theories, the hypotheses will be confined to particular subrealms of the theory. Thus, it is most likely that specific constituents of the theory — such as models, laws, and assumptions — will be subject to test, rejection, or modification (Stegmüller 1976).

Proper tests of hypotheses derived from the more abstract components of theory will always need some translation. Model parameters, initial and boundary conditions, and scale and scope are all important aspects that must be specified in translating from abstract components to workable hypotheses. Hypotheses may be restricted to the scope of the theory or may probe beyond its current scope. In cases where the hypotheses are tested beyond the scope of the theory that spawned them, it is domain that is being tested, although the hypothesis may appear to be about mechanism.

Examples of hypotheses are nearly limitless in ecology. Unfortunately, listing them in isolation of the theory and the set of observable phenomena that stimulated them cannot suggest their full importance. One testable hypothesis is the statement that the temperature of the densely hairy leaves of a desert shrub will be lower than the temperature of leaves from the same plant that have been shorn of their pubescence (Ehleringer and Mooney 1978). This hypothesis is derived from the general theory of physiological ecology, which has developed largely around the responses and characteristics of plants in environments considered to be severe. The hypothesis is founded on the thermal energy budget law (see the section titled "Laws"), which abstracts key features of thermal, radiative, and transpirational relationships of plants and information about a particular environment. It is also based on data and generalizations about the operating temperature ranges of plants. Furthermore, the hypothesis draws on the laws of physics when it links the presence of hairs, light reflection, and their effects on the energy absorbed by tissues. The specific model from which the hypothesis derives is that pubescence or hairiness on plants reduces the absorption of short wave radiation from the sun. This means that the leaf has less thermal load to dissipate to stay within an operating temperature range. Since the most common mechanism of heat reduction in leaves is transpiration, this means that a hairy desert shrub will require less water — which is of course usually limiting in deserts — to stay cool than would one without hair. The hypothesis, which proved to be correct (Ehleringer and Mooney 1978), thus derives from a set of observations about limiting and stress factors in desert environments and the relationship of various components of the energy budget.

In the realm of evolution, it might be hypothesized that an organismal characteristic thought to be subject to natural selection is heritable and genetically variable. This second example points out, first, that the "hypothesis" is actually a simple compound of two separate testable propositions. The first of the two prongs of the hypothesis is that the characteristic of interest exhibits genetically based variation in the population, and the second is that it is heritable. Falsification of compound hypotheses would not necessarily indicate which component had failed. Compound hypotheses, except for the most simple (of which this one is a transparent case), must be decomposed before they can be unambiguously tested, either by falsification or confirmation. This example also shows that the test of the component hypothesis dealing with genetic variability is not, as in the case of leaf hairiness, experimental. Carefully designed and conducted observations and measurements of the population would be adequate to test that subhypothesis. An experiment would likely be required to test unequivocally the component hypothesis concerning heritability.

An example of a hypothesis in the realm of ecosystem ecology is the hypothesis that nitrogen availability in a particular system is controlled via uptake by higher plants. Contemporary knowledge of nitrogen dynamics suggests that microbes also determine nitrogen availability in soils. Therefore, this apparently simple, directly testable hypothesis is a cryptic compound hypothesis that begs an alternative. At the least, microbial activity would have to be considered to be a boundary condition in the test of higher plant control of nitrogen availability by uptake. More constructively, the interaction of the two factors might be tested by devising a more sophisticated hypothesis, or perhaps a model, that makes predictions about the relationship of the three variables: plant uptake, microbial release or immobilization, and nitrogen availability.

In landscape ecology, an example of a hypothesis is the statement that increased landscape fragmentation is associated with lower species richness. This hypothesis is admittedly exploratory, and while clearly related to the theory of island biogeography, it is also associated with many other assumptions that must be specified in a terrestrial case. The basic theoretical structure is based on the assumption that immigration is reduced by small island size and distance from sources of colonists, while diversity on islands is reduced because extinction is increased by small population size. For that oceanic logic to be translated into specific predictions or hypotheses about terrestrial fragmentation, several model features must be put in place. First, what kind of species is of interest: core forest specialists, open site specialists, edge specialists, or generalists? How inhospitable is the nonforest matrix for the species of interest? Is there an edge effect that reduces effective island size? The point is not that island biogeography theory does not apply to terrestrial fragmentation but rather that framing testable hypotheses about it requires a clear model context. Application of island biogeographic hypotheses calls for pattern description in the new target habitat. This is to be expected in a novel or young scientific specialty (McDonnell and Pickett 1988). Generalization and the determination of causal explanation would come later in the landscape example and would be the subjects of later generations of hypotheses that evaluate the various assumptions discussed previously. An emerging area in which functional hypotheses are being tested in landscape ecology is in the area of boundary function (Cadenasso et al. 2003). By manipulating the architecture of the vegetation boundaries between forest and meadow habitats, the impact of the boundary on seed flux, nitrogen deposition from the atmosphere, and herbivory was discovered (Cadenasso and Pickett 2000, 2001, Weathers et al. 2001).

VI. Theory Frameworks and Structure

The various components of theory are not just randomly mixed constructs floating free in some disciplinary ether. They all have different functions and must be related to one another in some clear way. There are many ways to link ideas, ranging from logical entailment through deduction, to empirical condensation, to causal relationship, to mechanistic support, and finally to constraint. Some or all of these ways to link constructs and facts must appear in a theory if the theory is to be a useful touchstone for understanding.

A. Framework

A framework is the structure that relates all the other components of the theory. The importance of the framework is emphasized by the common reference to well-developed theories as "theoretical systems" (Suppe 1977a). The term "system" requires that the various parts of the theory be related to one another. The conceptual framework of a theory is a general model of the relationships of the various conceptual devices that constitute the theory (Box 3.2). It indicates, for example, what concepts are derived, how the assumptions are worked out, and what the conceptual inputs and explanatory or predictive outputs from the models are.

The definition of theory as a high-level — that is, general — system of conceptual constructs or devices to explain and understand ecological phenomena and systems is the only one that merits the term "theory" (Stegmüller 1976). Without a coherent system, the validity and cogency of the assumptions and, consequently, the accuracy of results cannot be evaluated (Murray 1986). In essence, theory cannot exist without a framework (Fig. 3.6). Conceptual constructs not explicitly connected to one another are often difficult to employ effectively (Gould 1984). According to Reiners (1986), an example of the limitation of integration by the lack of a framework appears in traditional ecosystem science, in which concerns of nutrient dynamics and energetics were pursued essentially independently. Recognizing such a gap is, however, a necessary step toward integration. Ecosystem ecology has begun to tackle this gap (Jones and Lawton 1995; Sterner and Elser 2002). For example, Elser (2003) argued that by connecting key concepts of ecosystem ecology, evolutionary biology and biochemistry, stoichiometric theory integrates biological information into a more coherent whole.

The contemporary semantic view of theories considers them as families of models (Brandon 1990, Lloyd 1988, Thompson 1989) with empirical content. The word "semantic" is used in the philosophical literature to differentiate the contemporary view from the classical "statement" view of theory. The definition in terms of a family of models suggests that the frameworks may be looser connections than the strict bonds of deduction. In fact, it is especially important to specify a framework in the contemporary functional view of theory as a conceptual device (Hacking 1983, Suppe 1977a) or a family of models (Thompson 1989), because the obvious rules of deduction are neither adequate nor applicable to the task. The failed positivist view of theory

Figure 3.6 A naïve theory builder ignores the framework. The framework defines how components fit together and what sort of structure is permitted. Definitions also are specific to the framework and may have no meaning or a different meaning outside the framework or in a different position within the framework.

as a series of statements assumed that theories were simple things whose components could be adequately joined by deduction alone. However, the connections may be functional ones, showing first how the domain is defined and how it constrains the other components of theory. The connections within theory may also show how definitions feed into or link the different models. They may show how specific tests can be framed, given the origin of hypotheses in abstract laws or generalizations. In other words, a variety of links must be brought to bear in the current view of theory. This is what a framework does.

1. What Theory Is Not

We give several examples of how the term "theory" might be misapplied. These examples also indicate why frameworks are so important to the use of theory. Some misapplications are relatively minor. One of these is Levins's (1968) definition of theory as "a cluster of models, together with their robust consequences." This definition does not flag the variety of components and structures that permit the models to be constructed and linked. Nor does it clarify the variety of forms that robust consequences can take.

Another popular but erroneous view of theory was promulgated by Rigler (1982). The so-called empirical theory (Rigler 1982) is not actually a system of conceptual constructs linked by a framework. Rather, it is a series of regression models that constitute, at best, confirmed generalizations rather than complete theories. Even if a larger number of regression models, or other statistical relationships, were successfully combined into an integrated network, almost all of the *explicit* assumptions of empirical theory would remain those of statistics. However, empirical theory tacitly uses a variety of ecological assumptions and concepts, but these are borrowed from the body of ecology and are applied implicitly without additional scrutiny. Empirical theory, as a matter of program, avoids dealing with the conceptual issues that underlie the use of such assumptions (Grene 1984). Moreover, the assumptions used by empirical theory come from disparate ecological theories and domains; without attention to their relationships, they cannot form a coherent system. The empirical theorists do, of course, have an important point to make in their call for a careful and operational measurement of pattern, but this good advice does not require the rest of the program they espouse which, in reality, is misleading in its narrowness.

We have already discussed a much more harmful concept of theory — the improper use of the term is the consideration of theory to be a hypothesis (Huszagh and Infante 1989). This error comes from an inductive method based on the statement view of theory. It is to be avoided. This misuse of the term "theory" occurs in public discourse and mars the understanding of science by the public. The vernacular usage "I have a theory," meaning "I propose" or "I suspect," should not corrupt the clear scientific use of the term "theory." For example, the "theory of evolution by natural selection" should not be confused with a popular notion or tentative hypothesis about how dinosaurs became extinct. Such an isolated proposition, being divorced from an explicit theory or line of scientific argumentation and evidence, is difficult to evaluate. Neither should the popular connotation of the term "theory" as something uncertain or undeveloped be brought into discussions of evolution. The theory of evolution in the strict scientific sense is one of the most successful and well-confirmed theories of science (Brandon 1981). It is neither a guess nor a collection of loose but narrow specific assertions about particular phylogenies. Likewise, a scientific theory like evolution must not be confused with popular metaphors such as "survival of the fittest," a phrase sometimes used sloppily even by scientists (Brandon 1990).

Using the term "theory" in the improper lay connotation of a guess also leads to confusion between legitimate scientific conclusions and the sometimes wild musings of individuals. This situation is especially a problem, for example, with the predictions of natural catastrophes sometimes made by self-proclaimed experts operating on the edges of the scientific community. Forecasting an earthquake on the New Madrid fault on a particular date is (so far) an example of unsupported speculation isolated from the successful core of geological theory. The New Madrid fault is one of the largest in eastern North America, and earthquakes centered there in 1811 were massive. So wishing to predict when the fault might slip again is a reasonable desire. However, the current status of geological science will not allow us to forecast the next quake there. Therefore, unsupported "expert" predictions are unfounded. Failure of any such wild disconnected statements leads the public to mistrust legitimate bodies of scientific theory, in part because of misuse of the term "theory." When some component of theory — such as a tentative generalization, proposed hypothesis, dynamic model, or untested prediction — is meant to be implied, such specific labels of the components should be used. When self-appointed experts make wild speculations, unconnected to established or explicit theory, these should be challenged (Bauer 1992).

B. Frameworks and Hierarchy

Theories are connected to external conceptual frameworks, as well as possess their own internal frameworks. Indeed, less general theories are often subsets of, or related to, more general theories (Fig. 3.7; Rohrlich 1987). Considering them in isolation may compromise their utility. Too little attention has been paid to the internal and external frameworks of ecological theory (cf. Lewis 1982). Note, however, that the theories of biology do not need to have explicitly nested within them the theories of chemistry and physics. We do not reawaken here the resolved problem of reduction of biology to physical sciences (Hull 1974, Mayr 1982, Sober 1984). There is no need to reduce biological theories to the terms of theories of physics, although biology must not contradict the physics of low-velocity systems.

Because theories must account for generality and allow for specific application, theoretical frameworks are likely to be nested hierarchies. A nested hierarchy means that components in upper levels of the hierarchy are made up of lower level components. Nested hierarchies constituting frameworks of theories must be considered distinct from hierarchies generated by scale of observation or frequency of rate process (e.g., Allen 1998, O'Neill et al. 1986). Levins's (1966) statement that different models accounting for generality, precision, or realism be nested to permit full understanding implies a hierarchical framework. General models would be higher

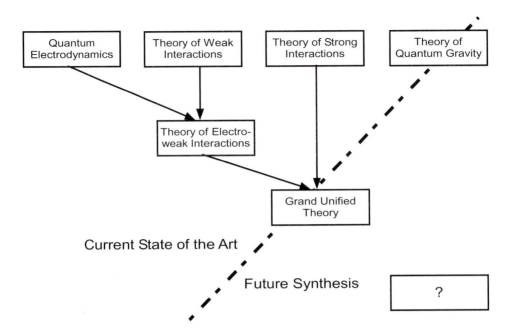

Figure 3.7   Hierarchical structure of major theories in modern physics. The empty box indicates the hoped-for unification; the dashed line indicates the boundary between firm and formulating theories. From Rohrlich
(1987). Copyright Cambridge University Press. Reprinted with permission of Cambridge University
Press.

level components of theory, with models nested within them that allowed for either precision or
for realism.

The hierarchical nature of theories permits different levels of theoretical systems to perform
different functions. Pattee (1973) noted that the laws of motion and equations of constraint on
these laws reside on adjacent hierarchical levels, as did the phenomenon of student mischief and
its constraint (Chapter 2). Such a hierarchical arrangement allows the first principles of mechanics to be stated without reference to their causal mechanisms (Miller 1987). This means, for
instance, that the laws of motion do not require theories of atomic or molecular phenomena to
be correctly formulated for the domain of low object velocities. The laws of motion exemplify a
dynamical theory, a type that can describe or predict the behavior of a single level (Pattee 1973).
Such theories are often supplanted by or supplemented with mechanistic multilevel theories
(e.g., Cohen 1985). The supplementation of the gas laws by statistical mechanics is an example.
The lower level statistical theory explains the gas laws but does not replace them at the macroscopic level. Within biology, a hierarchical structure of models relating to different levels of
organization is provided by Schoener's (1986b) rigorous reduction of certain models of community ecology to models of population ecology. Hierarchical interpretations of other specific
theories are offered by Thorpe (1974), Williams (1984), Campbell (1974a), Beckner (1974), and
Rohrlich (1987).

Different degrees of generality are represented by different hierarchical levels of the theoretical
framework. Therefore, hierarchical frameworks permit scientists to judge the generality of conclusions from a particular study or component of theory. As mentioned before, the net-effects
models of succession (Connell et al. 1987) operate on a very general level, but the site-specific

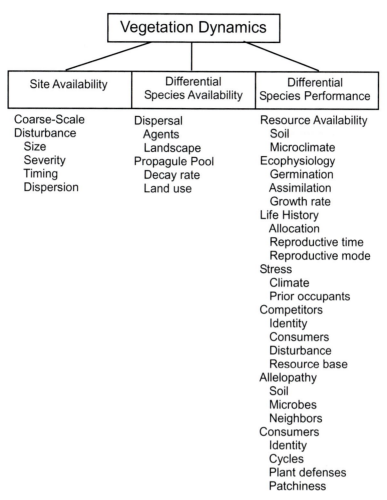

Figure 3.8 Hierarchical framework for a theory of vegetation or plant community dynamics. The top level indicates the general phenomenon of change in community structure or composition. The second level indicates the three major areas of processes involved. Each of the three processes is, in turn, driven by particular interactions and conditions. Modified from Pickett et al. (1987).

models or experiments on successional mechanisms require a lower level, more specific domain (Fig. 3.8). Generality of mechanistic models or conclusions about mechanisms can only apply to a particular level in the hierarchy (Allen and Hoekstra 1992, Kolasa and Pickett 1989, Pattee 1973, Salmon 1984) after other phenomena at that level, or at relevant lower levels of the hierarchy, have been evaluated or accounted for (Passioura 1979).

C. Frameworks and Open Systems

Frameworks anchor translation modes and therefore permit the use of theory to generate understanding. Translation between the general, abstract content of theory and the specifics of the material world is key to the dialogue between conceptual constructs and observable phenomena that constitutes understanding. We noted earlier that to make the abstractions and idealizations inherent in any general theory susceptible to test and application, specific submodels, subtheories, operational definitions, and parameterizations must be interposed between them and the reality of the environments in which the theory is to be applied.

The need for translation modes, connected to frameworks, emphasizes the empirical openness of theoretical systems in ecology. Theories confront observable phenomena and are modified as a result. The view of theory and its components that we have presented is appropriate for open systems of understanding. Of course, one key way in which theory is part of an open system is that observable phenomena can be absorbed into the theory as facts and as confirmed generalizations when shared properties are ascertained, as indicated earlier in the chapter. Erroneous components can be discarded from a theory or refined in it. In presenting the view of theories as open systems, we have been motivated by the view of understanding as a match between observed phenomena and some conceptual device. This perspective is akin to Boulding's (1964) view of science that involves an image of certain phenomena, the inference or expectation derived from the image, and the message that confirms or denies the image. This system has the advantage of being interactive and flexible. Note that it does not necessitate the theoretical versus empirical dichotomy assumed by logical positivism and its discredited descendant philosophical models.

Such an open system of understanding contrasts with closed systems of understanding (Danto 1989). The classical axiomatic approach to theory is the epitome of a closed system. Axiomatic systems necessarily begin with axioms and definitions. Rules of inference or logic permit consequences (or theorems) to be deduced from the axioms and definitions. No aspect of a closed axiomatic system need approximate observed phenomena. Geometry provides one of the best examples of a closed system. The axioms and theorems of a geometric system do not have to be those of our intuitive, nearly Euclidian world. The match of geometry and the everyday world is not required. In fact, point, or line, or a plane does not pretend to represent anything in the observable physical world. If one were to evaluate the understanding afforded by closed systems such as those of geometry, it might be based on internal consistency, deductive power, and completeness. Such systems have been ideals that philosophers have used as models for theory analyses of empirical sciences (Danto 1989). The philosophical holdover from such ideal closed systems appears in demands for deduction and prediction as defining criteria of science. Contemporary philosophers of science have discarded these criteria based on study of what empirical science actually does (Boyd et al. 1991).

It is clear that the open system of understanding is successful and appropriate to contemporary empirical science and that it entails a broad conception of theory such as that presented here. The interaction of the various components of theory produces the image; the interaction of theory with observed phenomena constitutes the message by which the image and its expectations are evaluated (Boulding 1964). The framework is an important tool for using theory and relating it to observable phenomena and, thus, for supporting an open system of understanding.

D. Domain

Each theory has a scope or domain in which it applies (Loehle 1987a, Mayr 1996, Suppe 1977a). By now, this is no surprise. But we close our discussion of the components of theory with this component because it is so crucial to the productive use of theory and so important to the goal of increasing integration. The degree of generality of a theory is judged by the breadth of its domain. The three aspects of domain noted in the literature are (1) the phenomena or concepts addressed by the theory, (2) the level(s) of organization to which the theory applies (Edelman 1974, Grene 1997, Thorpe 1974), and (3) the spatial and temporal scale(s) it addresses. Strictly, domain represents the partitioning of observable phenomena into those that will or will not be addressed by the conceptual constructs at hand. However, we suggest that, for convenience, domain be considered a part of theory so it will not be overlooked. This is akin to considering the definite article a part of the noun in gender-based languages. This maneuver is simply a mnemonic device to prevent domain from being neglected. Neglect of domain, as well as the failure to determine whether a theory is congruent in domain to the set of observable phenomena with which it is being compared, is a serious and common error. An example of the error of incongruent domains already mentioned is a domain of net effects versus an application to mechanism in competition theory.

As an example of domain of a theory, we cite island biogeography. The theory was constructed to apply to true, or oceanic, islands and to account for the appearance and disappearance of species as active components of the communities of the islands. The theory incorporated the processes of dispersal and of species extinction based on biotic interactions or rarity. The theory was expanded even in its focus on true islands by the realization of the importance of the former connections of land bridge islands to source areas (Castle 2001). This insight led to the inclusion of the concept of relaxation in species richness. Thus, land bridge islands were recognized as special cases of islands, on which the rates of extinction and colonization differ from islands of equivalent size that have never been connected to the mainland. Later applications of the theory — for example, to mountaintops, to other continental island analogs, or to individual plants or populations as islands — were successful to the extent that the assumptions of effective insularity and the relevance of the central processes of invasion, interaction, and extinction in a homogeneous environment were appropriate (Haila 1986). This example shows the importance of knowing the domain of the theory in applying it to novel situations and the ability to recognize subsets of the universe of application in which new causes or a different mix of the causes might operate.

The domain of the entire science of biology has sometimes been dealt with as a controversial subject. How universal is the domain of a theory or of a whole science? Mayr (1996) provided an entertaining example concerning life in general when he exposes the faulty nature of the argument that biology is not a universal science, as physics is claimed to be, because it does not project beyond Earth. Mayr pointed out that biology covers the entire known empirical domain of life phenomena — therefore it is as universal as physics is. It is important to specify domain and to evaluate claims about how broad or narrow the domain of a theory is.

Domain can appear in many places within a theory. Note that certain aspects of the domain might be flagged in the assumptions of the theory, whereas other aspects are encoded in the models or in other components. Domain is determined or inferred from all these reasonable sources in a theory, which is why we have discussed it last.

VII. Conclusions and Prospects

Our survey of the anatomy of theory has indicated that theory is itself a high-level conceptual construct or system, composed of more specialized conceptual constructs that interact with one another. Thus, theory has both richness and depth of components. Some of the components are entirely conceptually based, whereas others have a substantial empirical content. Some of the components are simple, whereas others are highly derived. Each of the components of theory performs a distinct function in establishing the larger conceptual construct against which observable phenomena are compared to generate understanding. The degree to which all the components are present in a theory is the criterion used to evaluate its completeness. We infer from the exploration of the richness of theoretical components that the individual components can themselves change. Such change in components results from the dialogue between phenomena and the conceptual constructs. We also conclude that the relationship among components is embodied in the framework of a theory. Frameworks are characterized by their integration of components, their openness, and their nested hierarchical structure. Hierarchical frameworks permit ecologists to work with (1) the various levels of organization within their science, (2) the structure of a particular theory, and (3) the relationships of a focal theory to more inclusive and more specialized theories. The analysis of frameworks suggests that the relationships among components of a theory can change. Such change, like that of specific components, is driven by the dialogue between phenomena and the theory. How, exactly, theories can change and the dimensions along which change in theory can be cast will be dealt with in Chapter 4. Changes in the relationships among components of theory will be portrayed in Chapters 4 through 6.

4

The Ontogeny of Theory

"Scientific truths are tentative and partial, and subject to continual revision and refinement, but as we tinker with truth in science-amending here, augmenting there — we always keep our ear attuned to the timbre of the web."
Raymo 1991:145

I. Overview

Theory is a system composed of diverse but interrelated components. The broad goal of scientific understanding in a subject area relies on the whole theoretical system for that subject. Consequently, the status of understanding depends on the status of theory. Theory can change over time due to changes in its structure and content. Specifically, theory can change along three axes: (1) completeness: components can be added; (2) development: components can become better worked out; and (3) integration: connections among components can become better articulated. Together, these changes constitute the ontogeny of theory and can be combined to define theory maturation and, thus, to help identify a mature theory. Ontogeny and theory maturation result in differences within a theory over time. How a theory can be used depends on its ontogenic state at a particular time. Improper use of theory can be prevented by knowing the developmental stage a theory represents.

II. Why Theory Change Is Important

Theories are not often static or fixed structures. They change in complex ways as science develops. Biology's master theory, the theory of evolution, provides an excellent example of how a theory can change with time. Besides the value of the example itself, there is delicious irony in the fact that the theory of evolution has evolved. The central domain has remained constant in its specification of objects, dynamics, phenomena, and population-based mechanisms. The theory has always sought to explain patterns of descent and mechanisms of modification in organisms. However, the broader domain, objectives, content, and structure of the theory have all been expanded in the time since the basic theory was originally proposed near the middle of the 19th century. Its mechanistic component has grown vastly, first by means of Mendelian genetics and later by the inclusion of molecular genetics. Important aspects of these genetic expansions recognize the role of mutation, the genetics of development, and transgenic inheritance. These expansions indicate that there are many mechanisms of genetic change. Evolutionary theory has also expanded to incorporate aspects of the newer discipline of population biology. This addition to evolutionary theory has highlighted the significance of the spatial and age structures of populations. Another important change in evolutionary theory is that the original mechanism of natural selection has been supplemented by the related but sometimes countervailing mechanisms of sexual selection and kin selection. Similarly, whether the same evolutionary mechanisms operate on different temporal and spatial scales or on different biological units has been explored. For example, the phenomena of punctuated equilibria and species selection have been proposed to explain patterns of stasis within lineages within the fossil record. More recently, gene selection within organisms has been explored. The contemporary expanded theory of evolution also recognizes that selection is constrained by various organismal and ecological contingencies. As a result, selection can be countered or shaped by aspects of organism development and anatomy as well as specific historical, ecological, and geological events. It is critical for biologists and the public to realize that the debates and controversies in the scientific community concerning the expansions to the basic domain of evolutionary theory are healthy and do not represent a dismissal of evolutionary theory (Futuyma and Moreno 1988). In fact, the changes are just the sort that would be expected in a vigorous, well-confirmed theory (Eldridge 1985).

The kinds of changes we have shown for evolutionary theory are not uncommon. Ecological theories change too, and knowing how these changes occur, and what they mean, is important. Because ecological understanding depends on theory, the state of theory will constrain understanding. Because ecological integration also depends on the status of theory, it too can be limited by theory development. Failure to recognize how theory changes or, indeed, that theory changes at all can be an impediment to furthering ecological understanding and integration. The classical philosophy of science, with its focus on complete and highly developed theories, may trick scientists into missing the importance of theory change. Neglecting theory change can lead to reliance on a theory that is incomplete or immature. Several failings may arise from using an immature theory. Many such failures are caused by the absence of components or of integrating links in the theory. More unfortunate, an immature theory may be prematurely rejected or entirely ignored when it is subjected to an overdesigned test at an early developmental stage. The relationship of theory change to testing is explored next.

A. Ontogeny and Testing

A practical reason to be aware of theory change is that theories in an early ontogenic phase may not permit an appropriate test. For example, theories in which components are absent or not well developed may not generate sound hypotheses — that is, the auxiliary hypotheses and assumptions needed for a fair test may be unavailable in a young theory. For example, in landscape ecology, theory is in its initial stages and has yet to develop the models of mechanism or constraint that link it to other levels of organization. How changes in landscape configuration and structure might actually control the fluxes of organisms, information, and matter is not yet well articulated (Cadenasso et al. 2003). Therefore, tests of landscape processes are often restricted to the level of pattern and do not expose potential mechanisms or discriminate among alternatives. The upshot is that testing hypotheses from incomplete theories must proceed with caution. One must be exceptionally careful not to throw out the baby with the bathwater, however: discarding an undeveloped theory because it does not yet admit to sophisticated testing does not mean that the theory has no value, nor does it mean that the theory will be rejected once it has developed to the point of admitting sophisticated testing.

Integration of a theory is also important in supporting its tests. The existence of a framework is critical for the successful use of any theory. A framework exposes the connections within the subject matter. Frameworks show what the logical structure is and what the causal interactions are in the theory. In other words, the framework shows the pathways of entailment and implications of the interactions. Knowing these connections allows researchers to interpret the confirmations or refutations correctly. A hypothesis and the data used to test it cannot be validated if the logical and causal connections that lead one to propose the hypothesis and the associated justification of the measurements are not clear. A refutation or a confirmation that arises from an incomplete or poorly framed prediction is not valid since the correct relationship between the predictions and the remainder of the theory is uncertain. For example, the widely known hypotheses or conjectures of Odum (1969) on trends in ecosystem succession were presented without reference to their basis or foundations, which made it difficult to develop and evaluate the theory.

Testing general theories also requires translation of their abstract and idealized models, laws, and concepts to the real world. Therefore, the elaboration of a framework to assist in constructing and using translation modes is a necessary prerequisite to testing a theory. We have already mentioned the models of Connell and Slatyer (1977) as exemplifying the lack of connections to a framework that differentiated net effects from causes (Chapter 3). Similarly, in plant apparency theory (Feeny 1976), the lack of rules indicating how one could distinguish plants that were likely to be found from those that were unlikely to be found by herbivores, made testing the theory difficult and perhaps led to premature rejection of the thesis (Grubb 1992). A more recent challenge to find appropriate modes of testing theory has arisen with the formulation of several general models of biodiversity derived from assumptions about distribution of individuals in homogeneous landscapes (e.g., He and Legendre 2002, Hubbell 2001). McGill (2003), who has analyzed the conditions necessary for successful tests of theory, emphasized the need to identify specific expectations. This requires researchers to translate the model into observable and measurable variables at appropriate scales. Selecting appropriate scales is also an element of translation. For example, the models in question predict species-area relationships but, as Rosenzweig (1995) has shown, the exact slope of species-area curves depends on whether they are derived from intercontinental data, island biota, or nested continental habitats. Each model must thus be provided with additional assumptions appropriate for producing predictions specific for these different scales before it can be sensibly tested.

Integration is an important factor in the successful use of theories. The philosophical truism that most theories are born refuted (Hacking 1983) indicates the critical and fundamental requirement for integration of a theory. If a theory is tested in a premature, or unintegrated form that does not take advantage of knowledge from other theoretical realms, that may lead to its inappropriate refutation. For using a theory, the following question arises about integration: Does failure of data to fit the predictions of a theory always lead to rejection of the theory? Simply put, the answer is no, because theory sometimes can change to account for discrepancies. Such change is appropriate and not a mere case of ad hoc desperation. For example, in island biogeography theory, failure to find clean deterministic relationships between island area and species richness led ecologists to discover the roles of history and habitat heterogeneity (Abbott 1980). Note that such changes cannot be idiosyncratic but must be consistent with other accepted theories and confirmed components of the present theory, and they must be derived logically or connected causally to the remainder of the theory (Quine and Ullian 1978). Because of the requirement that revisions or expansions of a theory be connected and consistent, the failure of a test of a theory can actually stimulate further development of the theory or its components. A good example of such development is offered by optimum foraging theory. Since the 1970s, the theoretical tradition and empirical tradition converged to form a more embracing and integrated framework (Perry and Pianka 1997).

Of course, repeated accumulation of sound refutations can lead to the rejection of a theory — if its internal structure is clear. An integrated theory that fails many tests will be more readily replaced if research strengthens an alternative theory. The combination of accumulated tests that argue against a theory along with the rise of an alternative theory (or theories) will require the rejection of a mature theory. For example, the demise of the view that plant secondary metabolites were waste products was abetted by the alternative view that they had defensive functions (Fraenkel 1959).

The complexity of issues and actual difficulties encountered during the process of designing tests of a theory should not lead to the erroneous conclusions that ecological theory is untestable as claimed by Peters (1991). The error has been repeated more recently by Smith (2000). The view that ecological theory is untestable arises from a simplistic view of what constitutes a test and completely ignores the fact that a theoretical postulate has to be translated by specifying the system, scale, test constraints, method of analysis, and other parameters before a reliable test can be executed and interpreted. Similarly, the fact that theory may change is not a sufficient reason to abandon theory in general and to replace it with a naïve reliance on empirical relationships alone.

B. Scope and Refutation

Theory scope and the potential for rejection are intimately linked to the ideas about theory revision presented earlier. Theories of broad scope, or domain, are less likely to be rejected by one or a very few negative tests than are narrow theories. This assertion assumes, of course, that such a theory is well developed. We expect narrow theories to have specific implications that apply only under narrowly defined conditions. Thus, the relevant universe of discourse is limited for narrow theories. However, a theory with a broad scope will more likely be divided into subdomains, each with well-developed models, assumptions, and translation modes to deal with special cases. Therefore, "exceptions to the rule" may be dealt with by specialized models. In a more narrow theory, or a particular model, the exception to the rule is fatal. For example, a focused equilibrium theory of community niche partitioning, embodied in the "broken stick" model (MacArthur 1972), was found not to fit the actual distribution of abundances in nature. The assumptions of sharp distinctions between niches and of the homogeneity of environments apparently defeated its application (Kingsland 1985). Yet broader theoretical approaches, still centered on competition and resource partitioning, maintain a major role in community ecology (Tilman 1982). Again, accumulated instances of refutation of some key underpinning or convincing refutations that cannot be accommodated by the growth or subdivision of a theory do require its rejection. A good example of theory growth is provided by the neutral theory of biodiversity (Chave 2004); tests of both assumptions and predictions failed either thoroughly or at certain stages of theory development. The assumption of equivalence among individuals or species met with a crushing amount of evidence to the contrary. However, a prediction that communities must necessarily diverge led to the modifications of the theory that brought it in agreement with the empirical evidence (Chave 2004).

C. Conceptual Refinements

In Chapters 2 and 3, we established that ecology, like all contemporary sciences, involves an open system of understanding. Openness means that the state of understanding is established by means of a dialogue between the observable phenomena and the conceptual constructs that are the discipline's theories. Openness also means that understanding can change as a result of the dialogue. Understanding can change as a result of expanding the phenomena available to the science through new technology, new scales of observation and measurement, new concepts, new approaches to manipulation and experiment, and, especially in ecology, the accumulating observations of rare events (Weatherhead 1986) and observations of new environments or habitats.

In addition to the empirical sources of novelty in understanding listed here, understanding can also change as a result of the ontogenic change in the family of conceptual constructs against which nature is compared. Ontogeny refers here, as it does in organisms, to developmental change over time. The best use of theory in ecology must account for the fact that the conceptual constructs can change independently of observation because of internal logical changes and the development of new concepts or clarification of existing concepts. In ecology, with its long history of concepts, this last avenue for improvement, conceptual clarification, is especially important. The open system of understanding in ecology requires us to exploit both the dialogue between theory and phenomena and the internal refinement of the conceptual devices.

Loehle (1987b) has described how theories must be in a certain state of maturity before they can be tested effectively. For example, optimum foraging theory with its precise language and formal models can be challenged experimentally in a variety of ways, including its many predictions and assumptions (Nonacs 1993). In contrast, ecosystem theory, which is still struggling with refining its vocabulary and methodology for developing expectations, does not invite tests as effectively, or such efforts do not meet with wide acceptance (e.g., Odum 1990, O'Neill 2001). Sometimes, the conceptual refinement may be difficult to notice and may represent a shift in interpretation of a single concept. Ives (2005) argued that this has been the situation with the recent increase in diversity-stability-focused research. Specifically, he believes that the meaning of stability, for all practical purposes, has drifted so that it now stands for lack of or only a small amount of variability. Such redefining of stability, from a primary focus on equilibrium models to patterns of variability, opened opportunities for great amounts of data to bear on the question and attracted the attention of ecologists who normally were less interested in the intricacies of differential equation modeling.

It is interesting to note that McCann (2005) provides a different though complementary view of evolution of the diversity-stability question. In the simplest terms, he emphasizes the changing relationship between the two concepts (Castle 2005). One might view such differences as indicative of theoretical immaturity, but whatever the judgment, the current situation illustrates how theory content and theory perception by practicing ecologists influence theory development and set the stage for further efforts.

Because of the complexities of theory change, discussed previously, to best use theory in advancing understanding, we must be able to assess the degree of development of a particular theory. Patience is a virtue in the application of theory. Premature rejection of a theory may deprive a field of a useful tool for integration and unification. Giving up on the development of a pretheoretic notion can be a mistake. Because scientists are trained to be critical and because they are generally deeply influenced by the hoary philosophy of logical positivism and its falsificationist descendants, the risk of ill-advised acceptance of a young theory seems less of a danger than throwing the theoretical "baby" out with the empirical gray water from the bath. Rather, we should become better at assessing the status and needs of theory and discerning when a developing theory can engage in the rigorous discourse with reality.

III. How Theories Change

The three dimensions of theory change are completeness, development (Box 4.1, Fig. 4.1), and integration. Completeness refers to the richness of components that are actually present in a theory. There is a specific roster of components (Chapter 3; Box 3.2) that have different jobs in

BOX 4.1 Axes of Theory Change

Completeness. Assessed by the degree to which the roster of theory components is filled.

Development. Assessed by whether a theory is functionally robust; well-developed theory is characterized by high degrees of exactitude and empirical certainty, along with clearly derived conceptual constructs.

Integration. Integration of theory components is reflected in dependence among components and connectedness between them.

Stages of Theory Change. Pretheoretic: Represented by rudimentary notions only. Intuitive: Simple and fundamental components present. Consolidating: Derived conceptual devices emerging. Empirical-interactive: Generalizations, laws, and hypotheses present and amenable to legitimate test. Confirmed or rejected: Judgment by the scientific community of adequacy and strength of tests (confirmatory or falsifactory) of the theory; confirmed theories often permit practical application.

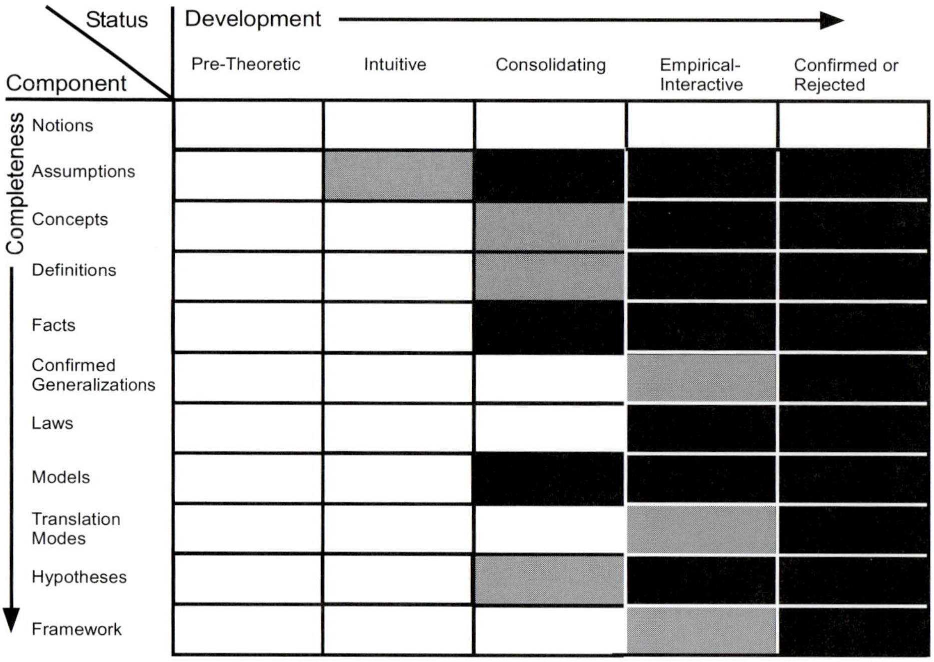

Figure 4.1 Two axes of theory maturity. Theories can change through the addition of components, which is indexed as completeness, along the *y* axis of the figure. The *x* axis applies labels to the varying degrees of development and suggests the functions that theory may perform at differing states of completeness. The degree of shading represents the change in individual components leading to increased refinement and precision. The third dimension of theory change, integration, is not illustrated in this figure.

a theory. Theories that are more complete can do more of what is required of a theory. Therefore, complete theories can more successfully abstract, idealize, generalize, unify, causally explain, and predict than those that are less complete. To accomplish these roles in science, theories must have basic conceptual devices, empirical content, derived conceptual devices, a statement of their limits and structure, and components that permit the explicit dialogue with observables.

Development refers to the degree of complexity or derivativeness of components of a theory. Components of theory that are better developed are explicit, clearly derived from other components of theory, empirically rich, and more thoroughly worked out than poorly developed components. They are also effectively linked to a specified domain.

Finally, the degree of connectedness of components in theories changes. Because theories are conceptual *systems*, the significance and utility of the separate components of a theory depend on some or all of the remaining components of the theory. Connectedness can be referred to as integration within a theory. The framework of the theory is the vehicle by which such internal integration is expressed. We will expand on the definitions and implications of the three axes of theory change in the following sections.

A. Theory Completeness

Richness of components is a hallmark of mature theories. We will present the components in their idealized order of accumulation as a theory develops over time. This is the same order found in Chapter 3, in which the order was taken to represent the increasing complexity of components. Once a theory begins to take shape and to be used, it will often become clear that existing components must be replaced or refined, or additions to certain kinds of previously existing components must be made. In other words, ecologists will often have to return to aspects of the theory that had been previously established to make additions or changes. An example of such change appears in the domain of succession theory, which for a long time focused on species replacement but evolved to accommodate shifts in community architecture as well (Luken 1990; Pickett and Cadenasso 2005). Similarly, plant defense theories initially focused on evolutionary reasons for the presence of secondary metabolites (Coley et al. 1985, Feeny 1976, Fraenkel 1959) but then expanded to encompass causes of change in ecological time (Bryant et al. 1983).

1. Domain

The first explicit component of theory that must be specified is its domain. Identifying what the domain comprises is not always easy. Within ecology, determining the domain of a theory requires familiarity with a wide variety of concepts and facts ranging across a wide spectrum of specialties. So while mastering the constituent concepts of any scientific proposition is difficult, it is especially so in ecology (Haila 1986). However, any theory must be founded on a clear specification of the universe of discourse. Without a stated domain, it will not be clear to what the theory applies, nor how the universe might expand, contract, or subdivide as the theory is refined and tested. For example, succession theory as first codified by Clements (1916) focused on a coarse scale that was appropriate to entire biomes. An early refinement in domain was to divide the climate-dependent focus on biomes into variation based on substrate type, which could better predict actual successional trajectories in local environments (Whittaker 1951). Similarly, developments in plant defense theory have refined the domain based on plant growth rates (e.g., fast versus slow growers; Coley et al. 1985), because expectations for these categories of plants are quite different.

One of the most exciting developments in ecology is the expansion of the domain of some of its theories. For example, ecosystem theory was developed initially for "natural" systems of plants and animals. Recently, the ecosystem concept has expanded to include human structures, institutions, and human-driven processes. Although the originator of the concept encouraged the expansion to humans (Tansley 1935), the incorporation of humans and the social, infrastructural, and interactive processes they affect has only been fully accommodated in the past few years (Machlis et al. 1997). Treating humans and their institutions as components of the ecosystem is a challenging and promising frontier. This expansion of domain shows one of the ways in which ecological theories can develop.

Using the history of island biogeography theory, Castle (2001) showed that discoveries in ecology are routinely made by extending models to new domains even when initial extension is somewhat risky. Such expansion of a theoretical domain occurred in island biogeography. The theory began with a focus on oceanic islands with a completely hostile matrix separating them from the source of colonists. Applying this model to terrestrial systems required ecologists to recognize analogs to islands that might exist on continents, such as ponds, mountaintops, or new successional habitat. In addition, they came to recognize that the matrix surrounding island analogs would not always be entirely hostile to migrating colonists. The most complete conversion of island biogeography might be said to be patch dynamics, in which a terrestrial habitat is conceived as a mosaic of patches, each of which interacts with neighboring and distant patches to different degrees. In this stage of the transformation of island biogeography to the land, there is essentially no matrix, and all patches have the capacity to interact. This transformation from the sea to the land has been laid out by Pickett and Rogers (1997).

2. Assumptions

Once the domain is clear, the assumptions about connections, phenomena, dynamics, and entities can be specified. Assumptions, as we observed in Chapter 3, are those guesses about how the system is built, how the components interact, and what sorts of dynamics can occur. The importance of assumptions in the development of theory is shown by cases in which assumptions have not been initially recognized. The failure of simple Lotka-Volterra models led Hutchinson to develop the previously tacit assumptions of no social interactions or a lagged effect of density (Kingsland 1985). Likewise, tacit assumptions that plant secondary metabolites were mostly biologically active have been challenged and shown to be incorrect, requiring substantial modifications to defense theories (Jones and Firn 1991). As a theory develops, assumptions may be further elaborated and teased apart in the specific models. Revisiting assumptions may be a productive job at any stage of theoretical development (Box 4.2).

3. Concepts

Concepts appropriate to the domain and consistent with the assumptions can be specified once these fundamental components are in place. We expect simple concepts to be specified earlier in the history of a theory than derived concepts, because derived concepts are composites of the simpler concepts. Population growth is a simple concept. More complex is the concept of lagged population growth. In contrast to the situation in which a simple concept is the first one developed in a theory, sometimes complex concepts are the starting point. Subsequently, they break down into more specific components. The evolution of the concept of succession followed this pattern. First, the concept referred to directional change in community composition. Later, it was revised to recognize that changes in architecture could be accounted for by the same mechanisms as change in composition. Hence, the concept of succession became broader and more inclusive as the theory developed.

BOX 4.2 Exercise: Theory Change

Understanding and articulating theory assumptions often lead to meaningful tests and advancement in a field. Here we suggest that the reader attempts to do the following:

1. Choose a theory that one is or wishes to be familiar with
2. Make a list of assumptions already stated by the authors of the theory or their followers
3. Attempt to identify situations where the theory or its assumptions do not apply
4. Recasts those lacunae (gaps) as implicit assumptions
5. Make a list of possible consequences should these assumptions turn out to be false or inapplicable
6. Publish these exciting results

---

4. Definitions

Before concepts can be combined into models and laws, it may be necessary to define limits to certain dynamics and phenomena, to define the units of quantitative relationships, and to define other parameters that are necessary to the development of laws and models. In community dynamics, the basic concept that communities change through time was refined by recognizing that different environments might produce at different rates. Therefore, a useful definition was the discrimination of primary and secondary successions. Different patterns and rates of succession could be better accommodated by recognizing the two types of succession. A similar case appears in plant defense theory. Defining and distinguishing induced resistance (a net effect on herbivores), chemical induction (an underlying cause of change in plant resistance), and induced defense (an inferred evolutionary consequence) (Karban and Myers 1989) have been particularly valuable in clarifying the structure of the theory as a whole. More recently, Loehle (2004) argued that inadequate vocabulary (i.e., "definitions") impedes advances in the most challenging problem of ecology, that of ecological complexity. Among examples, Loehle cited such an obvious problem as one resulting from the poor definition of study subjects as incommensurability among studies ostensibly measuring the same thing. O'Neill (2001) raised the same issue in his assessment of the ecosystem concept and theory. These examples all point to the important functional origin of definitions based on their role in a theory. Definitions are not merely the labels of concepts, nor are they arbitrary semantic decrees. Rather, they arise from the operational needs and structures of the theory.

5. Facts

Facts are the confirmable observations that become part of a theory. Facts may exist prior to the existence of a specific theory or they might be discovered when the developing theory stimulates the search for them. Facts predating a theory are illustrated by Thoreau's (1863) observations on the order of invasion of pines and oaks on abandoned farmlands and the role of animals in the process. These observations suggest a succession of communities, yet it would take another 30 years for the theory to be scientifically articulated. In contrast, observation of trophic efficiencies depended on the existence of trophic dynamic theory (Lindemann 1942). Before Lindemann (1942) proposed the theory, there were no "facts" of trophic efficiencies.

Importantly, facts may come to have a particular meaning in a certain theory compared with their meaning and significance in a different theory. For instance, "disturbance" was only an initiator of coarse-scale succession in classical theory (Clements 1916). Now, however, it is seen as a potential fine-scale organizing factor in late successional communities as well (Pickett and White 1985). It may well be that as a theory changes, existing facts are discarded as irrelevant or are demoted in importance. The facts that supported "ontogeny recapitulates phylogeny" have been retired as the theory that gave them a place was dismantled.

Facts can stimulate the development of other components in theory. The existence of confirmed generalizations follows the accumulation of a sufficiently supportive body of facts. Although Hutchinsonian ratios of size in pairs of competing species have not survived subsequent analysis (Kingsland 1985), they were initially proposed based on an accumulation of cases (Hutchinson 1959). Confirmed generalizations can serve as hypotheses themselves, or they can suggest new hypotheses. Thus, empirically derived hypotheses are the next components of theory that are expected to proliferate. We will not say more about hypotheses here, because their role is functional rather than structural in theory. That function has been laid out already in Chapter 3. We therefore move on to the role of laws as derived conceptual devices in theory.

6. Laws

Laws, the quantitative or verbal statements of relationship, constraint, or dynamics in a theory, are likely to come next in the elaboration of a theory. Laws are more complex than the previously available components of theory because they have an internal logical structure that must be derived or an internal causal structure that must be empirically supported. For example, the law of natural selection arose by analogy with other kinds of selection (i.e., plant and animal breeding) and by combining that concept with Malthusian limitation and a statistical view of variation in organisms (Thagard 1992). The law of natural selection is a highly derived construct. The analogous law of vegetation dynamics combines observations of change in the community realm with generalizations about physiological and architectural differences between plants and the accumulating observations about the role of various environmental and organismal factors. Most of the laws of ecology listed in Chapter 3 behave in a similar way, even if the theory they are part of may not be fully mature because of insufficient integration, incomplete assumptions, or vagueness of some of the conceptual constructs (discussed later).

7. Models

Models, like laws, usually represent a highly developed state of theory because models have an internal structure and require simpler components of theory for their construction. Perhaps not surprisingly, specific models may exist before the complex theories they represent. Thus, models exemplify a complex component of theory that often precedes other components. Once a model is developed, the need to elaborate a complete theoretical system may become apparent. Investigators then work to identify the domain, extract hidden assumptions, and ensure that the concepts are rigorously conceived and stated. In this way, models can serve as the stimuli for the generation of a whole theory. However, it is critical not to confuse a model with a whole theory. The fundamental components discussed earlier and the derived components necessary to apply models and laws must be present in a complete theory. The growth of island biogeography theory in the ecological literature represents this scenario. The central graphical models (MacArthur and Wilson 1967) preceded elaboration of a broad theory that could account for habitat heterogeneity, land bridge islands, island analogs on continents, and terrestrial patch dynamics (Pickett and Rogers 1997).

8. Translation Modes

Translation modes are especially critical aspects of theory that must be developed to apply the models, laws, concepts, and abstract definitions of a theory to specific field or laboratory situations. The specific form and structure of translation modes depend on the laws and models to be applied and on the situation in which they are to be applied. Translation can be by means of more specific causal or dynamic models or relationships that require specific definitions of boundary and initial conditions. For example, landscape ecology (Forman and Godron 1986) was first formulated using the human-centered perspective of landscapes. To apply the theory to phenomena associated with animal populations, translations had to be done from human-determined patch types to the patch types actually discriminated by animals (Merriam 1984, Opdam et al. 1984).

Hypotheses in ecology, the specific testable expectations derived from a theory, are likely to rest in a web of translation modes rather than emerge, pure and simple, from complex models and laws. Even when the basic assumptions of a theory serve as hypotheses, care must be taken in applying them to field or laboratory. We have already mentioned the operational translation of the net effects hypotheses of Connell and Slatyer (1977) to models that accounted for resource levels and community structure (Armesto and Pickett 1985). We present another example of translation using a law introduced in Chapter 3: *In the absence of evolution, any habitat will asymptotically approach an equilibrium number of species, with the equilibrium value being a positive function of habitat size, as long as immigration and extinction rates remain constant.* The translation rules for this law need to address a number of potential issues. One is the implicit assumption that the habitat does not change, which, of course, cannot strictly be met in any except an idealized system. Indeed, in any natural habitat, changing species composition inevitably implies changing of the habitat attributes. Thus, the translation rules must involve some criteria of degree to which the habitat change is permitted for the test of the law. By the same token, other translation rules must circumscribe the taxonomic boundaries of the test, the relationship between the habitat and the surrounding matrix of communities that is the source of immigrants, the significance of rare events such as large disturbance or impact of a new immigrant predator or competitor on community structure, and a time scale at which the asymptote can realistically be found.

9. Frameworks

Frameworks are the logical and causal structures of the theory as a whole. They differ from models, which are logical and causal pictures of specific phenomena that the theory treats. Because frameworks pull together an entire theory, most components of a theory must be present. The components must be assignable to a clear place in the framework to display their connections to and implications for the remaining components of the theory. Evolutionary theory is composed of models of population genetics, modes of selection, and levels of selection, whose connections have been summarized by Lloyd (1988) and Thompson (1989). In the case of contemporary evolutionary theory, the framework shows how the outputs from the genetic theory are the inputs to the selection theory (Thompson 1989). The greater completeness of contemporary evolutionary theory is illustrated by the more extensive framework relative to Darwin's original (Thagard 1992). The aspects emphasized in the contemporary theory are just those that embody mechanisms that were unavailable to Darwin.

Because the components of theory acquire meaning and significance in relation to one another, a framework is needed to express those relationships. Translation modes then become identifiable subsets of the framework that allow the abstractions and idealizations of the assumptions, concepts, models, generalizations, and laws to be applied under particular circumstances. The theory of punctuated equilibrium is a novel translation mode of the broader theory of evolution that establishes a new relationship between evolutionary theory and the fossil record. With the new translation mode in place, the long periods of stasis in the fossil record, punctuated by episodes of change, are seen as significant (Eldridge 1985, Gould 2002).

The axis of theory change we have just described is the degree of completeness. Completeness can be assessed as the proportion of the ideal richness of theory that actually exists at any time. This axis refers to a particular theory through time and represents growing complexity and derivativeness of components of theory. We emphasize that the meaning of different components of theory cannot be given without consideration of the position of theory along this axis of development. The meaning, or context, of the components of theory will change as more components are added and as those components themselves change. We have presented the sequence for adding components in its ideal form, as though a theory were built from scratch. Because theory change is so often haphazard and reflects an amalgam of different theories, and even initially empirical pursuits (e.g., Hacking 1983), we have noted some instances in which more complex or highly derived components can arise before simpler components in a particular theory.

The key idea about theory development is that several specific jobs must be done in any theory and that some component(s) must be present in a complete theory to do all those jobs (Fig. 4.2). The basic conceptual devices provide the core ideas of the theory. Components of theory that help accomplish this task are assumptions, definitions, and simple concepts. The second major job of theory is to organize and stimulate empirical observations. Empirical content of theories includes observations, facts, and confirmed generalizations. Combining empirical content with the simple conceptual devices or combining two or more simple conceptual devices yields derived conceptual devices. Laws and models are the derived conceptual devices encountered in biological theories. The job of providing the limits and structure of the theory are accomplished through translation modes, the framework, and the domain. All of these tasks and components yield testable hypotheses, which accomplish the dialogue with the observable phenomena in the domain of the theory. Of course, depending on the content and structure of the various components, there may be some compensation of one for others. For instance, models in some theories may play the same role that laws play in another.

Several cautions and consequences apply to growing theory completeness. Clearly, in elaborating or using a theory, ecologists must be alert for missing components. However, because theories often draw on other theories for components, ecologists must also be aware of problems in such transfers. The significance and meaning of a law, relationship, or concept may change when it is inserted into a new or different theory. Because established components of theory may have well-known names, the transfer may conserve the preexisting name, but the *role* of a transferred component may be entirely different in the recipient theory than in the source theory. For instance, in ecology, the physical laws of conservation of matter and energy become assumptions in ecosystem theory (Reiners 1986). The transfer of such laws from physics to the incipient theory of ecosystem ecology does not mean that the new theory now has its own laws. Indeed, one goal of ecosystem theory is to discover the laws that govern fluxes in ecosystems (Sterner and Elser 2002, Walker 1991); clearly, the physical laws of the conservation of matter and energy need to be reinterpreted in recognition of the biotic components of any ecosystem. The ecological laws of ecosystem function may be more likely embodied in generalizations about stoichiometry and elemental balances, for example, than thermodynamics. These stoichiometric relationships combine the principles of chemistry with the constraints of physiology in plants and animals and the differences that likely attend different modes of consumption. Herbivores, omnivores, and detritivores each face different stoichiometric constraints and opportunities. One trick to avoid

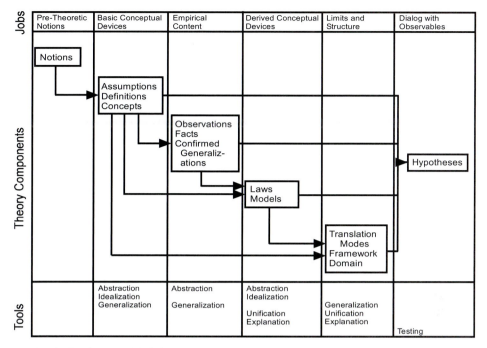

Figure 4.2 Jobs for theory and their idealized developmental sequence. The top row of the figure incorporates the broad jobs of theory, the middle portion indicates development of the components, and the bottom row includes the tools for building and using theory. Theory emerges from pretheoretic notions and begins to form when basic conceptual devices are available. An empirical content is adopted or stimulated early in the development of theory and, in conjunction with the basic conceptual devices, contributes to the generation of derived conceptual devices. The limits and structure of a theory may be proposed early in the development of the theory, but they are only fully formulated when most of the other components of theory are present and available to be linked. The dialogue with observables takes place throughout the development of the theory, but a certain high degree of development is required for legitimate attempts at falsification. Although all actual theory change should represent all the jobs, the order in which they appear may differ in reality from the ideal order shown here.

the confusion of adopting laws and generalizations from one theory into another is to employ the term "principle." A "principle" refers to any of the derived or conceptually based components of theory and is therefore useful in reducing confusion in transferring components among theories. Similarly, concepts such as entropy, imported from physics to ecosystem theory (Odum 1983), may acquire a different meaning and interpretation in their new context. The most appropriate approach to this problem is to be clear about what role a tool plays in the theory under discussion. What is a fact, concept, assumption, law, or generalization in the focal theory?

B. Theory Development

The individual components of theory can be well or little developed. In the discussion of theory completeness, derivativeness and complexity were cited as attributes of various components of theory. Here, we explore more completely how components of theory develop and, hence, a second aspect of how entire theories change.

1. Exactitude

Exactitude or precision is an important attribute of components of theory. Exactitude varies in proportion to the answers to questions such as the following: Are the components of theory explicit or hidden? Are the assumptions hidden in the structure of the models and of the entire theory, or are they clearly identified and stated? Are all aspects of the domain stated? That is, are the objects, relationships, scale, and dynamics that are of concern clear? How specific are the statements, models, and generalizations of the theory? Specificity is an appropriate criterion for verbal, graphical, qualitative, or quantitative models.

Clarity of terms is an important aspect of exactitude. Quantitative terms possess the advantage of having to be clearly specified. Any symbol used to represent a quantitative term of a model necessarily must be defined in terms of other concepts or definitions in the theory at hand or in theories that are taken as background. The same is true of verbal and graphical models. Unfortunately, verbal and graphical relationships may suffer from imprecision. Ecology is conducted throughout the world in living languages. Although this has the advantage of making the discipline accessible to a great number of practitioners and users, it has the serious disadvantage of permitting confusion. The vernacular meanings and connotations of terms are insidious in ecological usage (Pickett and Cadenasso 2002).

Ecologists must make diligent efforts to state explicitly and clearly what the apparently vernacular terms they use mean in the more demanding context of their science, and they must not be misled by street connotations. For example, the term "disturbance" in ecology is sometimes inappropriately laden with its negative vernacular connotation, so surprise is expressed when some part or all of an ecological system responds positively to a natural disturbance. The development of disturbance theory has proceeded from using "disturbance" in a technical sense to mean some destructive impact on a community to a more useful definition that is independent of the nature of the response, as well as being applicable to various levels of organization other than the community (Pickett et al. 1989, Rykiel 1985). Taking "disturbance" as a general technical term that refers to the disruption of a defined system portrayed in an explicit model solves the problem. The first task in addressing disturbance is to elaborate a structural or functional model of the system of interest over a specific spatial and temporal scope. Without that, disturbance cannot be used in a clear technical sense.

Such important efforts are not "just semantics." To conduct a productive discourse, the scientific meaning of the term in the relevant scientific context must be accepted by anyone who wishes to engage in the discourse. Without specificity, the apparent focus of discourse is illusory, and either agreement or disagreement is merely a mirage. Imprecise words usually mean muddled thought. Furthermore, as long as terms are encumbered with subjective impressions, relevant only to a specific research project or, worse, to narrow cultural or personal experience, ecology can have no firm generality. Therefore, to enhance integration in ecology, fear of abstraction from specific cases and disdain of specificity in terminology must be overcome. It might be easier if science were still conducted in classical Latin!

2. Empirical Certainty

Empirical certainty is another of the features of theory development. It may be evaluated by questions such as the following: Are the facts well confirmed? Are the empirical generalizations well confirmed and statistically evaluated? Note that there is a place for both probabilistic and deterministic relationships in well-developed theories (Humphreys 1989, Suppe 1977a). An example of differing degrees of empirical certainty derives from forest regeneration. The conceptual model of forest dynamics, driven by alternating periods of canopy opening, canopy closure, and species exclusion, is fairly well confirmed (Oliver and Larson 1990) but is only a tentative hypothesis for herbaceous communities (Armesto et al. 1991).

3. Applicability

Applicability to the observable phenomena is another aspect of development. In well-developed theories, the abstractions and idealizations that stand at the core of the theory are supplied with translation modes, allowing operational hypotheses and operational models to stand between them and the reality of the field, glasshouse, or laboratory. An area in which translation modes are still developing is conservation biology (e.g., Ehrlich 1989). In this area, the mathematical theories of population genetics and of structured populations come together to suggest a useful concept, the minimum viable population size. Translation modes to apply this parameter require specific data on the populations to be protected, as well as more complete demographic theory and a theory of habitat structure (Shafer 1990).

Determination of applicability may require involved empirical research when the scope and modes of application are not obvious or inferable a priori. A process akin to trial and error may be needed to accumulate knowledge necessary to determine the requirements for a successful application of the theory.

4. Derivativeness

Derivativeness refers to the feature of development that requires all the implications of individual components and their relationships to be worked out. Here we can cite the modern use of the term "analysis," which implies thorough working out of implications and ramifications rather than reduction *ad infinitum* (Gruden 1990). Compound concepts and models especially need such analysis and exploration of their inferences. Horn et al. (1989) used an analytical approach to simulation models of forest dynamics. They introduced a new level of generality into such models by dividing the universe of discourse. The general phenomena are modes of tree mortality and gap filling (Fig. 4.3). The universe is a two-dimensional one, defined by whether or not a gap is created by the mortality of a tree and whether or not a tree requires a gap to establish. This seemingly simple stroke was an analytical breakthrough for gap-focused simulation models.

Theory development is stimulated and driven by the dialogue between theory and the world (Chapter 2; Fig. 2.1). This dialogue is the basis for changing the components of theory, which may lead to refinement and replacement. Recall that ecological theory is an open system of understanding, and the mutability and development of theory are a key attribute of the success of an open system. Static, or closed, theory could not respond to the dialogue. The relationship of theory change to the dialogue between conceptual constructs and observable phenomena is explored further in Chapter 5.

Even when attention is aimed primarily at conceptual clarification, the goal is improved ability to confront the observable world. But in practical terms, there can be an internal dialogue within theory that influences its development. The refinement of conceptual devices can be stimulated by examining their place in a theory, initially independent of how the theory performs in the dialogue with the world. An example is the clarification of the concepts of disturbance and stress based on an underlying assumption that a persistent system (of whatever level or scale) will have some minimal structure and function. The concept of disturbance is clarified by limiting it to disruption of the physical or structural aspects of the system, whereas stress is linked to a reduction in functional performance of the system. Stress on one level can lead to physical disruption of higher levels, just as disturbance can cause stress elsewhere in a nested hierarchy (Kolasa and Pickett 1989). Refinement of the individual components of theory is a sound basis for changing relationships among components, as well as for replacing components. One can well imagine that discovering an irreparable internal inconsistency in a theory would lead to its rejection. Indeed, internal consistency is one of the hallmarks of successful theory (Kuhn 1977).

C. Integration within Theory

The increasing interrelationship among parts of a theory is the final of the three major axes of theory development. This axis evaluates the connectedness, dependency, and integration among the components of a theory. That the various parts of a theory be present is not sufficient; they must be functionally connected to one another, as required by the existence of a framework. The parts of any theory depend on one another for their meaning and utility. For example, models are not useful without the assumptions that guide their application, nor are laws concerning dynamics useful without their empirically based constants or the definition of initial or boundary conditions. Similarly, isolated observations and facts are literally meaningless. Therefore, the establishment of a framework is needed to specify the relationships among theoretical components, so they can be worked with and changed if necessary.

A framework must show the pathways of derivation and inference within the theory. In this sense, the framework serves as a model of the entire theory, since it specifies the parts, the relationships among parts, and the output of the dynamics of the conceptual system. If the things a framework must do sound like the same things that a model of a physical or ecological system must do, that is appropriate. Because frameworks are models of a conceptual system, they have the same jobs that other models do.

An example of a theoretical framework is that found in succession theory. The founders of succession theory recognized a large array of causal factors (Clements 1916, cf. Miles 1979). However, how these factors might be related, which ones should be excluded from idealizations,

Figure 4.3 A forest gap viewed from below. This single canopy gap, the result of an experiment in the Kane Experimental Forest of the USDA Forest Service, was created by cutting an individual black cherry tree in a forest approximately 60 years since clear cut logging. The diameter of the gap is approximately 5 m. Gaps are often critical resources for the release of suppressed seedlings and saplings of both pioneer and later successional tree species (cf. Collins and Pickett 1988). Gaps created by older, larger trees can be considerably larger. The size of gap also depends on the severity of the wind event that creates them.

which ones could modify idealized trajectories of systems, which ones operated as community-level mechanisms versus landscape-level constraint, and so on were not recognized before a hierarchical framework became available (Luken 1990, Pickett et al. 1987, Pickett and McDonnell 1989). This framework, along with contributing models, indicates how to unify the various kinds of approaches to vegetation dynamics that had been pursued independently (Connell et al. 1987, Huston and Smith 1987, Tilman 1988).

Certain changes are expected in a theoretical framework as the theory as a whole is elaborated and used. The domain will become refined, and subdomains will be added as the framework develops. For example, dividing community ecology into specialized realms depending on the nature of the resources, the degree of influence by other communities, and life histories of the predominant organisms shows the subdivision of a domain that otherwise invites confusion when broad generalizations are attempted or general hypotheses tested (Schoener 1986b). As a theory develops, the translation modes also change as the framework develops and as premature attempts at testing are found to have failed because of inappropriate domain or specific assumptions that prove to be faulty. The error is often in applying too broad a domain. In the case of community theory, models or experiments must translate from the abstract to the specific, based on resource level and supply patterns, the openness of the community to invasion, and the life histories of the species involved. When these factors are taken into account, the experimental designs become appropriate translations. In such translation and application, considerations of scale and level of organization are especially important.

IV. Theory Maturity

The maturity of theories can be assessed as a composite of the three axes of change discussed so far. Mature theories are complete, have well-developed individual components, and have well-integrated components (cf. Lakatos 1970). The complex axis of theory maturity reflects the various jobs within theory (Fig. 4.2). All mature theories must have a basic foundation of conceptual devices, empirical content, derived conceptual devices, specified limits, internal structure, and hypotheses as testable output.

The complex axis of theory maturity can be divided into phases for convenience reflecting the status of a theory. The continuum stretches from intuitive to empirical-interactive. Its phases are (1) pretheoretic, (2) intuitive, (3) consolidating, (4) empirical-interactive, and (5) confirmed or rejected. The phases indicate the possible stances of theory in the dialogue with observable phenomena. Only a mature theory can be judged to be confirmed or rejected. As noted earlier, extreme caution must be exercised in rejecting an immature theory — according to Lakatos (1970), such a theory should be sheltered against radical tests. On the other hand, the likelihood of rejecting a mature theory outright is low, assuming it is consistent with other theories and has been confirmed empirically in the past. A well-confirmed, mature theory is more likely to be revised in content or domain than rejected by a single legitimate test. Perhaps mature theories are more likely to be overthrown by a paradigm shift than by straightforward empirical disconfirmation. Such a shift would involve an entirely new view and approach to the world that could replace a well-confirmed, mature theory of large domain. The existence of an alternative theory enhances the overthrow of one that has accumulated negative tests.

An interesting and current case of theory maturation in ecology involves the neutral theory of biodiversity mentioned earlier and originally proposed by Hubbell (1979) as a neutral model but later developed to include a richer roster of more clearly defined components. Initially, this theory led to a prediction that communities should diverge in space and time (Chave 2004). Empirical tests of this prediction were negative, which posed a dilemma of whether to reject the theory or change its assumptions. Subsequent developments led to theory improvement, from clarification of domain, rejection of erroneous assumptions, identification of implicit assumptions (e.g., inadmissibility of density-dependence of population processes), to reformulation of its mathematical structure.

We can also return to the example with which we opened the chapter to illustrate the growth of a theory. The ferment in evolutionary biology is quite healthy and represents an expansion of domain to levels of organization other than the population and to mechanisms that Darwin did not even dream of (Eldridge 1985). Yet the theory remains robust and has great explanatory and predictive power. To be sure, its predictions are not the simple forecasts about the trajectory or exact branch point of a particular lineage demanded by naïve critics of the theory. Historical contingencies and existing evolved constraints are too great for that sort of forecast (Gould 1989, 2002). However, more fundamental predictions and confirmation, for example, that all life on Earth should share the same fundamental system of inheritance are quite successfully defended. The development of evolutionary theory by incorporation of such new principles as punctuated equilibrium and debate within the scientific community about mechanisms of selection above the individual level (Lloyd 1988) do not indicate a weakness of the theory. Rather, they are appropriate attempts to extend the domain and incorporate new causal mechanisms based on new observations and concepts. The misuse of these debates and developments by creationists and proponents of "intelligent design" in the attempt to discredit evolutionary theory is a transparently erroneous tactic.

The dialogue between theory and observable phenomena is strongly influenced by the ontogeny of theory (Box 4.3). Indeed, it depends on where in the three-dimensional space of

---

BOX 4.3 Theory Maturity and Interpretation of Data

When theory is immature, facts insufficient, framework missing, questions incomplete, domain vague, and assumptions implicit, even a good experiment and deduction may lead to serious mistakes.

> But I have learned by this handicraft-operation, that all Vegetables do immediately, and materially proceed out of the Element of water only. For I took an Earthen Vessel, in which I put 200 pounds of Earth that had been dried in a Furnace, which I moystened with Rain-water, and I implanted therein the Trunk or Stem of a Willow Tree, weighing five pounds; and at length, five years being finished, the Tree sprung from thence, did weigh 169 pounds, and about three ounces: But I moystened the Earthen Vessel with Rain-water or distilled water (always when there was need) and it was large, and implanted into the Earth, and least the dust that flew about should be co-mingled with the Earth, I covered the lip or mouth of the Vessel, with an Iron-Plate covered with Tin, and easily passable with many holes. I computed not the weight of the leaves that fell off in the four Autumnes. At length, I again dried the Earth of the Vessel, and there were found the same 200 pounds, wanting about two ounces. Therefore 164 pounds of Wood, Barks, and Roots, arose out of water onely. — Jan Baptista van Helmont (Brussels, 1577; Vilvorde, 30 December, 1644)

Excerpts from Ortus medicinae, Id est, initia physicae inavidita. Progressus medicinae novus, in morborum, ultionem, ad vitam longam . . . (Amsterdam: Elzevir, 1648), translated by John Chandler (as Oriatricke, or Physick Refined, the common Errors therein Refuted . . . , London 1662, 1664) and reprinted in Henry M. Leicester and Herbert S. Klickstein, *A Source Book in Chemistry*, 1400–1900).

maturity — defined by completeness, development, and integration — a theory is located. The maturity of a theory therefore determines, in part, the state of the open system of understanding resulting from the dialogue between it and the relevant observable phenomena. All the changes in theories summarized by maturity constitute a directional ontogeny as theories become more complete, components are better developed, and the entire structure becomes better integrated. The changes in theory over time that can be depicted in the three dimensions of change define maturity. However, theories that have different objectives will differ even at the same ontogenic stage; this is reflected in the taxonomy of theories (Chapter 5).

V. Conclusions and Prospects

This chapter has considered how theories change. For theories to be most useful, some large proportion of the potential richness of components must be present. The most useful theories will incorporate explicit assumptions, clear domain, clear concepts and definitions, a body of fact, confirmed generalization, laws, models, a framework with translation modes, and hypotheses. Not only must some large proportion of the components of theory be present, but the individual components must be well developed for a theory to be maximally useful. Development refers to exactness, empirical certainty, applicability to observation, and the derivativeness of complex components. In addition, connections among components must be specified, since the components of theory gain meaning and utility only in the context of the whole theory. These three major sorts of change lead to maturity of theory.

   This chapter has described how theory changes as understanding is created and improved. Indeed, the structure and content of a theory must change if understanding is to grow. The next chapter considers how theories are classified based on their objectives and domain. In conjunction with the insights summarized in this chapter about theory change, the nature of kinds of theory will be integrated into an overview of the entire spectrum of theory dynamics.

5

The Taxonomy of Theory

"Metaphor may be the spark that ignites scientific understanding, but the expression of the flame soon leads into a fractured maze of specialized vocabularies."

Raymo 1991:179

I. Overview

The previous chapter showed that theories differ in maturity. In this chapter, we explore two other ways in which theories can differ. Differences in objectives and differences in domain can result in contrasting structures of theory. Differences in objective reflect the variety of tools that contributes to the dialogue between theory and observable phenomena. The kinds of conversations in the dialogue represent different specific objectives of science and result in different theories with specific emphases. The second major type of difference in theory reflects the differing foci or domains that theories can have. Different domains usually invoke different dynamics, phenomenology, and causes. The taxonomy of theory reflects the differences among the goals, the structures, and the contents of different theories. By extension, the taxonomy of theory shows the different modes of understanding that are available to ecologists. This chapter presents these sources of difference in theory and gives examples of theories or models illustrating the different kinds of theory.

II. The Bases of Taxonomy

There are two reasons, other than ontogeny or maturity, that theories may differ. First, theories can differ as a result of having different *objectives*. The objectives are set by emphasizing different aspects of the dialogue between conceptual constructs and observable phenomena. Because the dialogue can emphasize abstraction, idealization, generalization, unification, causal explanation, or prediction (Fig. 2.1), each of these tools of dialogue can serve as the primary focus for a theory. For example, theories may either *describe* or *explain* dynamics. Similarly, theories can be either abstract of concrete in focus. The list of possible strands in the dialogue suggests an equally rich array of theory types. However, this diversity can be organized. The different tools can be sorted into three general dimensions of contrast by which theories can be classified. Based on differences in objective, the three major classes of contrast between theories are (1) phenom-

Table 5-1 Kinds of Theory

Kind of theory or model		Reference
I. Instantaneous versus Historical		
Compositional	Evolutionary	Shapere 1974
Ahistorical	Historical	This text
II. Phenomenological versus Mechanistic		
Phenomenological	Mechanistic	Leary 1985
Dynamic	Mechanistic	Pattee 1973
III. Abstract versus Literal		
Axiomatic	Nonaxiomatic	Leary 1985
Formal	Factual	Leary 1985
Hypothetical	Empirical	This text
Principle	Constructive	Einstein, in Palter 1984
Normative	Descriptive	Stegmüller 1976
Abstract	Literal	Leary 1985
IV. General versus Specific		
Pure	Operational	Grant and Price 1981
General	Operational	Levins 1966
General	Specific	Leary 1985
Strategic	Tactical	Holling, in Yodzis 1989

The dichotomies show a variety of ways to characterize theory. Approximately synonymous characterizations are grouped. N.B. Categories III and IV are continua rather than strict dichotomies. Details can be found in the references cited.

enological versus mechanistic focus, (2) general versus specific, and (3) abstract versus literal (Table 5-1). These axes will be defined and exemplified later.

The second major way in which theories can differ is as a result of their having different *domains*. The specific domains of theories obviously call for different content, but the differences in domain can also call for different structure and organization of theory. For example, a clear difference exists between theories that focus on systems that can be explained and understood by considering the present state alone or a simple transformation to or from the present state and theories that focus on systems whose history has persistent, recurrent, or lagged effects on the trajectory of change. Such a contrast in domain results in theories being characterized as either instantaneous or historical. Another way to conceptualize this contrast is as equilibrium versus nonequilibrium. Note that the other three axes of theory (Table 5-1) can be subdivided into historical versus instantaneous. The four general classes of theory mark the endpoints of axes along which theories can be contrasted. The four axes are not mutually exclusive. Therefore, theories may be jointly characterized by several of the contrasts.

A. Instantaneous versus Historical Theories

The contrast between instantaneous and historical theories is determined by whether the current state of objects depends on their prior state or only on current interactions. Synonyms for this class of taxonomic contrast include ahistorical versus historical (cf. Gould 1989) and compositional versus evolutionary (Shapere 1974). We add equilibrium versus nonequilibrium, because in strict equilibrium theories the interactions are considered to be stationary, or the same through time (Facelli and Pickett 1990). Sometimes, for simplicity, ecologists analyze relationships as if they were the same. The driving forces may vary in strength, which causes change in the system, but the same forces continue to act in the system. In nonequilibrium theories, episodic and rare events may sometimes alter the dynamics of the system, and understanding the current state of a system is not sufficient to understand the trajectory through time. In other words, the details of the history matter in nonequilibrium and historical theories.

Equilibrium niche theory in community ecology is an example of an instantaneous theory, whereas evolution or succession theories represent, at least in part, theories that explicitly account for history of the entities. Niche theory is instantaneous in focus because the niche characteristics are considered givens that are then used to explain community structure and species richness. The history of how the community is assembled is irrelevant, and only the present degree of contrast between species is used to explain the coexistence of the species.

Historical theories assume that the order of events, the existence of indirect effects that work themselves out through time, or the occurrence of one or more key episodes in the past determines the current state of the system. These historical events can be as important as current status and interactions in explaining system structure or behavior. For example, succession theory recognizes the importance of legacies of the initiating disturbance in the contrast between primary and secondary succession.

Historical theories can use either an evolutionary approach or a simpler chronological approach to the role of past processes. The evolutionary approach casts explanations in terms of accumulated, heritable change. In contrast, the chronological approach posits persistent or echoing effects of past states of the system without accumulated heritable change. Both approaches can be generalized as a part of the newer theoretical structure that is emerging in the area of ecological resilience (Gunderson 2000). In this general view, it is the adaptability and ability of a system to adjust and change in response to major environmental events or shifts that allow the persistence or sustainability of the system to be understood. Note that this take on "resilience" is different from the equilibrium-based definition of speed or degree of return to a predisturbance state.

The evolutionary approach can employ an idealized model of potential optimization through natural selection, which then influences the heritable characteristics of organisms. In other words, the past events that generate the relevant organism characteristics are a response to natural selection. The evolutionary approach does not, however, require that a historical theory deal in optimal organismal structures or behaviors or that all features be interpreted as adaptive. Of course, the countervailing or filtering processes of evolution such as drift can play a role as well. The key insight is that when these events occur at a specific point in time but leave a genetic template that shapes organism response or function at a later time, history becomes key to understanding the system.

Other historical theories may also involve some ideal model. The existence of a central idealization such as optimality is a powerful comparative anchor for both theory and empirical study. Idealizations serve as points of reference for real or constrained dynamics of systems. In the case of succession theory, a central idealization in the classical theory was the climax, a state of compositional equilibrium that a community would reach in the absence of disturbance and secular environmental changes. Ultimately, because the classical idealization was so remote from the reality of the world (Botkin and Sobel 1975), an idealization focusing on an ideal *trajectory* of change in community composition and structure was substituted. The newer idealization is admittedly more complex, but it reflects species interactions with one another and with the local or changing external environment (Pickett and McDonnell 1989).

The important concept of "contingency" emerged as an insight from historical sciences, such as paleontology and evolution (Gould 1989). A useful and ecological analysis of the methodology of the discipline of history itself shows how apt historical approaches in fact are for science

(Gaddis 2002). Contingency simply means that the course of the dynamics or the particular trajectory followed by entities depends on external constraints that operated singularly or episodically at some point in the past, or on constraints embodied in the evolved or accumulated current status of the entities. The present array of animal phyla, based on body plans, is contingent on past accidents of mass extinction and survival. Importantly, the laws and processes embodied in the theory of evolution can still be used to explain the changes in the phylogenies that do exist and to predict the mechanisms that will interact in generating future phylogenies. One of the first realizations that the classical philosophy of science (Chapter 1) was incomplete resulted from its failure to consider biological systems that had persistent effects of history (Grene 1984). A cartoon illustration of the difference between instantaneous and historical theory can be achieved by imagining Newtonian mechanics where the acceleration and direction of movement of the object would depend not only on the force applied to it but also on the force that was applied before the current force. A theory that must consider such change in relationships may assume a different structure in response to this different demand.

B. Phenomenological versus Mechanistic Theories

The second axis along which theories can be contrasted is one of phenomenology versus mechanism (Leary 1985). A synonym for this continuum is dynamic versus mechanistic (Pattee 1973). The term "phenomenon" refers to anything that exists or happens — or, as the Greek root suggests, something that is shown or appears. Pattee (1973) conceived of phenomenon and mechanism as being hierarchically coupled, meaning that the phenomenon of interest is located on one level of organization whereas the mechanisms that contribute to that phenomenon are located on the next lower level. Here, mechanism refers to causes in general. For instance, if animal distribution is chosen as the phenomenon of interest, an important contributory mechanism will be the processes of behavior. Such a hierarchy is nested, but although it is necessary to go to a lower level, it is not necessary to examine the lowest possible level to provide an adequate mechanism to explain or predict the phenomenon of interest (e.g., Rosenberg 1985). Usually, one nested hierarchical level below the phenomenon is sufficient (Passioura 1979, Thornley 1980).

In constructing hierarchies of explanation like the one that Pattee (1973) suggests, the specific levels are different for each science and indeed even for different models within a discipline. It is therefore not necessary to proceed to the same level for all phenomena. For example, landscape ecological phenomena may often be explained by interactions at the level of the patch, while organization of a community is often explained by population characteristics and interactions. Population explanations, as the focal level, often draw on genetic differentiation at the lower level of individuals. Causal explanation requires a clearly enumerated roster of possible causal mechanisms, referred to as a causal repertoire (Miller 1987), as well as a hierarchical framework to show how the various causes can be nested within one another and how they all relate to the phenomenon of interest in a theory (Chapter 3).

Examples of phenomenological and mechanistic theories are the original theory of Mendelian genetics and that of molecular genetics, respectively. Mendel's theory was developed to describe the patterns of inheritance. Molecular genetics examines the suite of cellular and genetic processes of mechanisms of inheritance. Another example relates to the global patterns of species diversity described by models that are largely phenomenological in nature. These models involve correlations with time, productivity, and habitat heterogeneity (e.g., Quinn and Harrison 1988, Ritchie and Olff 1999). The phenomenological models may ultimately invite, and later come to include, mechanistic and historical explanations such as rescue effect in maintaining a population in an isolated or unsuitable location or after repeated fragmentation of a tropical rain forest through time.

An important extension to the hierarchical approach to explanatory theories proposed by Pattee (1973) is recognizing that cause also operates from *higher* levels of organization. That is, the phenomenon of interest may be constrained or shaped by coarser spatial scale or longer temporal scale events and processes than the scales on which the phenomenon exists (O'Neill et al. 1986). For example, to explain vegetation composition, the lower level causes might include competition and herbivory, while coarser scale, higher level causes might include climate and topography and their effect on the species pool available to a local community. Climate and topography are determined by such large and slowly changing phenomena as the Earth's orbital geometry, tectonics, and erosion, and so are slow and relatively persistent compared to the faster dynamics of plant community change. By comparison, the dynamics of competition and herbivory occur at the scale of roots absorbing ions or the travels of insects over the surface of a plant. They are smaller and faster than the community dynamics.

C. General versus Specific Theories

The third axis for classifying theories is general versus specific (Leary 1985). Synonyms include pure versus operational (Grant and Price 1981) and general versus operational (Levins 1966). Earlier, we showed a similar distinction applied to laws (Fig. 3.2). According to Levins, models and theories can be characterized as general, precise, or realistic. However, only two of the three characteristics can be maximized simultaneously, not all three. This unavoidable trade-off can be compensated for by nesting different models and by clustering models that deal with a subject (Levins 1966). The fact that theories are systems of conceptual constructs or families of models linked by a framework accommodates the nesting and clustering strategy quite well. Because frameworks of theories are often hierarchical, the nesting of various models that emphasize generality, precision, or realism is readily accomplished. Levins's (1966) insight about this fundamental trade-off between different characteristics of models is one reason any theory of more than local applicability will be more complete (Chapter 4) than any single model.

Examples of the contrast between generality and specificity in theory are shown by the contrasts of a theory for all succession versus one for secondary succession in mesic fields or a theory explaining the feeding of all herbivores based on plant nitrogen content versus one explaining feeding by specialist versus generalist herbivores based on secondary metabolites. The theoretical structure of landscape ecology compared to the more specific theory of boundary function and the models that operationalize those models is a further example (Box 5.1).

D. Abstract versus Literal Theories

The final axis is the contrast between abstract and literal theories (Leary 1985). Synonyms include axiomatic versus nonaxiomatic (Leary 1985), formal versus factual (Leary 1985), hypothetical versus empirical, principle versus constructive (Einstein, in Palter 1984), and normative versus descriptive (Stegmüller 1976). For example, optimal foraging theory is normative, indicating how a system will behave under certain specified ideal conditions. One does not expect a real system to actually behave in the ideal fashion. Rather, the ideal provides expectations against which deviations can be explained and referenced. Thus, the derivation and availability of these expectations are powerful tools for organizing knowledge required for understanding of the phenomenon. Predator/prey theory in general, compared with a model of moose and wolves in Isle Royale National Park, is a further example of abstract versus literal approaches to theory

---

BOX 5.1 Contrast between a General and a More Specific Theory

Aspects of a framework for a general theory of landscape ecology would include the high-level process that the theory addresses. This might be identified as "the creation, change, and flux across heterogeneous space." This high-level pattern-process pairing suggests the concern with the description, measurement, and explanation of spatial mosaics or three-dimensional fields of contrast. Component phenomena, which would be further expanded by specific models, would include (1) patch or gradient creation, (2) patch or gradient change, (3) structure and function of patch boundaries, and (4) nature and control of fluxes across the spatial mosaic or field. Of course, spatial and temporal scale must be specified for both the general and specific models that emerge from such a skeletal framework. The general theory of landscape is not fixed to any particular spatial or temporal scale. See Wiens (2001) and Pickett et al. (2000) for discussions of some high-level concepts of landscape ecology as an explanation of heterogeneity and its function.

A theory of boundaries can, for some purposes, be concieved of as a subtheory of landscape ecology theory. In any event, such a theory takes as its most general explanatory goal the concern with flux across patch boundaries or gradients. The more specific, yet still rather general, phenomena that must be addressed to understand boundary structure and function are (1) patch contrast, (2) nature of the flux, and (3) structure and processes within the boundary (Cadenasso 2003b). As in general landscape ecology theory, the domain of the general boundary theory is not limited to any particular scale. Time and space dimensions must be specified in the particular models constructed to address the issues identified in the general framework of boundary theory.

(Starfield and Bleloch 1986). Levins's (1966) trade-off and compensations, mentioned earlier, are relevant to this axis also.

III. Understanding and Diversity of Theory

We have seen that different characteristics of the dialogue between conceptual constructs and observable phenomena are associated with different objectives of theory. Each strand in the dialogue suggests a specific objective. The variety of theory objectives, therefore, has consequences for the use of theory. Each of the different kinds of theory will be used in different ways. For example, there are some theories whose domain and objectives simply do not call for a causal explanation, so such an approach cannot be supported by those theories unless new tools are added to them. Similarly, generalization is possible only when an empirical or constitutive theory (Table 5-1) has a sufficient body of fact and confirmed generalization to summarize and abstract that factual body of knowledge. An axiomatic or normative theory may help us to understand or predict the existence of some general relationships but, at the same time, it may be weak at addressing local specific needs. Such needs can only be satisfied after a roster of other devices and procedures — which may include translation modes, identification of system structure, and others discussed in earlier chapters — has been properly deployed.

Each of the kinds of theory will permit different aspects of the multifaceted dialogue between conceptual constructs and observed phenomena to be conducted or emphasized. This variety of emphases is valuable because the scientific process requires a combination of activities to carry on the dialogue between concept and reality. No single objective of theory takes precedence automatically. Each contributes to the dialogue, and each is appropriate to different choices of method suggested by the objects or interactions studied by a specialized area. Such a pluralistic view has great potential to alleviate unproductive and damaging debate about what single method of science is best, or which theory that emphasizes a particular specific tool of understanding best explains a phenomenon. Each is best for a particular choice of objective. The scientific method is more a braided stream than a single straight channel.

Different modes of understanding are reflected in the dichotomies between theory types (Table 5-1). There can be phenomenological understanding or mechanistic understanding. However, a unification of the two approaches may be possible. Such a unification can occur when, ultimately, we have a mechanistic understanding, based on sound and repeatable patterns of phenomena. Unification of observable phenomena can also occur via additional abstraction. For example, patterns in space such as fronts of migrating animals or distribution of burned patches and patterns in time such as fluctuation of measles infections may be found to represent cases in the broader class of chaotic phenomena. There may be instantaneous or historical understanding of an ecological phenomenon. Ultimately, we seek an integrated understanding of the existing states as they are entailed by any persistent effects of prior dynamics. There may be abstract or literal understanding of a phenomenon. Both are ultimately required, since abstract understanding allows the range of possibility to be explored, whereas literal understanding allows in-depth explanation or prediction in particular cases. There can be general understanding or operational understanding. Generality allows unification, prediction beyond the scope of existing observations, and an explanation of universal patterns. In contrast, specific and operational understanding allows the peculiarities of a particular case or trajectory to be teased apart. Ultimately, both the ability to project beyond the local and the ability to understand the local are needed. Fortunately, once theories mature, the possession of a nested hierarchical framework can accommodate a variety of models, laws, subdomains, and kinds and structures of generalizations, so the overall understanding permitted by the theory can possess aspects of all the different modes of understanding implicit in the taxonomy of theory (Table 5-1). In other words, what may first appear to be incompatible goals and structures can in fact operate together for a more complete understanding. This unification should be one of the major goals of a science. Understanding the axes of contrast between theories can actually help to suggest how to bridge the gaps between the extremes so that a more comprehensive integration can be constructed.

IV. Examples of the Classification of Theories and Models

Here we present several examples of how entire theories or important components of theories differ in their objectives and, hence, are classified differently. We consider first the case of the competitive exclusion principle. This generalization reflects a rich array of observations from biogeography along with observations on the structure of local communities. Furthermore, the principle is based on models of competitive overlap on a single resource axis (MacArthur 1972). The principle can be summarized in this way: In a uniform environment, species with similar resource demands will not coexist (Hardin 1960).

The complaints about the application of the competitive exclusion principle are many and well taken (cf. Murray 1986). However, it should not be considered an operational prediction of community theory. Rather, it is a central idealization (Box 5.2) within the theory having the form of a law (Box 3.2). In terms of the taxonomic categories we have presented (Table 5-1), the competitive exclusion principle is mainly of empirical origin; it is general rather than operational,

---

BOX 5.2 The Competitive Exclusion Principle

If two species occupy a homogeneous environment, and if those species have congruent niches, then they cannot coexist at equilibrium.

---

phenomenological rather than mechanistic, and ahistorical. Once these characteristics are recognized, the principle is seen as an abstracted empirical generalization that implies in its ideal features the conditions in nature under which coexistence might occur. Even at such a general level, the principle suggests several predictions: (1) increasing environmental heterogeneity increases species richness (as opposed to the assumption of environmental uniformity in the principle), and (2) increasing the number of environmental resource combinations increases the number of possible coexisting species (as opposed to the assumption that resources can be represented as a single, albeit complex axis). The principle also suggests that the openness of competitive systems might permit coexistence. This last prediction is in contrast to the tacit assumption by the competitive exclusion principle that the community of interest is a closed system to which migration is unavailable as an escape from extinction (Skellam 1951).

This analysis of how to use the competitive exclusion principle suggests that other important principles in ecology can benefit from such a taxonomic analysis. The example has suggested that it is important to know what components of theory a statement represents and what other components it draws on, especially if its assumptions are not usually stated. It further suggests that understanding the taxonomy of a theory or major component of theory can identify the tools in the dialogue between observables and conceptual constructs that will help apply the principle.

Another example of the utility of taxonomy of theory emerges from community ecology. The Lotka-Volterra models can, as suggested earlier, also exemplify several taxonomic features of theory. Recall that the Lotka-Volterra models are not the whole theory of community ecology. Rather, they are aspects of population dynamics theory and can be subsumed into community theory as idealizations about interactions (Schoener 1986a). These equations are normative — that is, they state what will happen under the ideal conditions of their explicit and tacit structural assumptions. The Lotka-Volterra models assume instantaneous and continuous reciprocal adjustment of populations, uniformity of environment, consistency of resource type, and the existence of a fixed carrying capacity. The models are phenomenological; the alphas represent, but do not specify, the mechanisms of competition. These models are not sensitive to the history of the competitive system. Many of their characteristics reflect the strategy of analogizing population theory with classical mechanics (Kingsland 1985, McIntosh 1985). Yet these same equations can drift into operational/specific types when they take on estimated parameters of population growth, reproductive delays, spatial diffusion, energy storage, size structure, and other features (e.g., Nee 1990).

Island biogeography, in contrast to the two examples just presented, is indeed an entire theory. It explains and predicts the number of species on oceanic islands, based on the processes of colonization and extinction. In turn, the rates of these two processes are constrained by distance from the source of colonists and the size of each island. The theory is operational because the variables are directly measurable. The theory is mechanistic at the level of entire islands because it specifies the role of extinction and colonization; however, the theory is not mechanistic at the level of populations because it does not explicitly invoke the autecological mechanisms of colonization and causes of extinction.

These original taxonomic limits of island biogeography theory have been superseded by subsequent developments. The original theory does not discriminate among species. Extensions such as the idea that certain species behave as supertramps do discriminate among species (Diamond 1974). The core of the theory did not accommodate history, but the extension embraces history by incorporating the idea that land bridge islands would result in relaxation from a higher than predicted species richness than oceanic islands of a comparable size and distance from sources. In addition, the theory was expanded to account for the heterogeneity of habitats and elevations on particular islands. Such heterogeneities on islands can modify the extinction rates. An additional domain expansion is the inclusion of terrestrial host plants as islands for assemblages of herbivorous insects (Janzen 1968) or lakes as islands (Keddy 1976). A broader theory of island ecology includes and permits alternative assumptions about dispersal (Abbott 1980). As island biogeography developed, it moved from being an abstract, general, ahistorical theory, to one that was operationalized in a wide variety of specific cases, to one that specified additional important mechanisms, and one that took account of history. Such changes typify theory development in ecology (Castle 2001). Island biogeography in its general form helped stimulate the development of landscape ecology in North America, and in its broadened form it expresses relationships that are the subject of the broader and still developing theory of landscape ecology.

A class of widely used models that simulate forest dynamics illustrates other characteristics that split theories into different taxonomic categories. Originally introduced as JABOWA by Botkin et al. (1972) and extended as FORET and a host of others (Shugart 1984), these "gap models" are based on patches within forests that represent the gaps created by the fall of mature canopy trees. Gaps are opened stochastically and are colonized on the basis of specified seed availability. Gap filling and the performance of species on the site are based on known autecological features of the species present and on the environmental changes, usually in light availability, resulting from the gap filling. Although this class of models is not a whole theory, it has proven to be exceptionally successful and generalizable in simulating forest dynamics when parameterized appropriately for a wide variety of biomes and disturbance regimes (Horn et al. 1989). The models do contain some necessary constants that are not mechanistic in origin. Ecosystem parameters have been incorporated, and competition for soil resources is now included as a basis for competition in gap filling (Coffin and Lauenroth 1990). Taxonomically, the original models were empirical, operational, and descriptive. The simulation models themselves make forecasts rather than predictions as we have defined them (Chapter 2). However, Horn et al. (1989), by developing a wider conceptual context for important inputs of species and disturbance types relevant to the models, began to deduce some novel expectations. They predicted, for instance, that removing any species other than one that produces gaps via mortality but does not require gaps for regeneration leaves a system that can still exhibit equilibrium. The prediction identifies gap producers that do not require gaps as keystones. This step led the gap models in the direction of generality. This development of the modeling approach makes it the core of a more abstract and generalizable conceptual system, while still maintaining the power of the core model as a simulation tool.

The examples of theories and components just discussed generate several insights. First, although theories may be of a primary or characteristic type, they can also have attributes of other types. This reflects the fact that the taxonomy of theory is based on four partially independent dimensions. Theories may also change type as they mature, especially if they add components that permit causal explanation. Such complex changes in theory depend on maturity and on objectives and will be explored more fully in Chapter 8.

V. Conclusions and Prospects

This chapter has considered how theories are classified (Table 5-1). Theories do a variety of jobs, including abstraction, idealization, generalization, causal explanation, and prediction. However, a specific theory may be capable of or may emphasize only one or a few of the whole array of objectives. The differences in objectives can be portrayed on four axes of taxonomy for theory: instantaneous versus historical, abstract versus literal, general versus specific, and phenomenological versus mechanistic. Both entire theories and their component models may be characterized along these four taxonomic axes.

We have treated the components contributing to the diversity of theory as separate. In Chapter 4, we explained how a theory can change through time, and so can be considered to have an ontogeny. In this chapter, we have shown that theories can differ depending on the objectives chosen for the dialogue between conceptual constructs and observable phenomena and on the nature of the domain. In the following sections of the book, we will gather together all the modes of change that theory can experience to emphasize the nature and utility of theory in permitting the growth of the open system of understanding in ecology, and we will illustrate some of the tactics and cautions in integrating theories and, hence, increasing ecological understanding.

Part III

From Theory to Integration
and Application

6

Fundamental Questions:
Changes in Understanding

"Delight in the unexpected is part of the lifeblood of science. Almost alone among belief systems, science welcomes the disturbingly new."
Raymo 1991:138

I. Overview

Enhancement of integration in ecology is the motivation for exploring the structure and use of theory. This goal leads us through a network of connected ideas, philosophical perspectives, and linked tools. We have presented understanding as the means of achieving and assessing integration and have shown that understanding relies on theory as one of its pillars. We described a model of understanding as an open system, resulting from the dialogue between conceptual constructs — or theory — and observable phenomena. We have, therefore, evaluated the role of theory in ecological understanding and have analyzed the anatomy, ontogeny, and taxonomy of theory.

However, we have yet to fully support the assertion that understanding advances integration. In this chapter, as a prelude to examining ecological integration in detail, we show how change in understanding is motivated within a specific domain in ecology. Understanding changes as a result of fundamental questions. Fundamental questions, in turn, arise from an awareness of the shortcomings of a theory or the need for a new theory in a domain. Fundamental questions are the most effective tools for advancing understanding because they address any one of five ways to improve theory: (1) adding components, (2) refining components, (3) rejecting components, (4) replacing components or entire theories, or (5) increasing the scope of theory. When several fundamental questions are competing for attention, they may be ranked according to their logical precedence, clarity, and potential to unify. Finally, we briefly consider the nature of fundamental questions as represented by the intuitive stroke of genius that leads to an entirely new insight.

II. Theory and Change in Understanding

The kinds of changes in a theory that lead to changes in understanding have been described as the ontogeny of theory (Chapter 4). In describing ontogeny, we have shown the wide variety of ways in which theory can change. Being aware of the ways in which theory can change is a key step in promoting that change. However, we have yet to see what drives the changes in theory. What kinds of insights or choices require ecologists to alter their theories and, consequently, to improve their understanding? To answer this question, we will start with a brief review of the structure of understanding. The relationship between systems of conceptual constructs that constitute theory and the observable phenomena that constitute the subject matter of a domain constitutes understanding. Understanding is a state generated by the process of general explanation. This, in turn, consists of testing, generalization, and causal explanation (Chapter 2).

The specification of a domain — that is, the suite of objects, kinds of relationships, relevant dynamics, and scale of phenomena to be addressed — is a critical step in achieving understanding (Chapter 3). Discussion of all these aspects of understanding has been restricted, so far, to a focus within a domain representing a particular subject area. We will continue that focus in this chapter. This focus within a domain prepares us for integration, however, because analogous changes drive the integration of distinct domains, which we will discuss in Chapters 7 and 8. Therefore, examining how understanding changes within a subject area introduces the larger task of integration. Examples of domains of understanding in ecology include ecophysiology, ecosystem energetics, community organization, competition theory, and landscape ecology (Fig. 1.4). In any particular subject area, a change in understanding requires some change in theory. What are the kinds of changes in theory that alter understanding?

A. Stimuli for Theory Change

There are five logically possible causes of change in theory. (1) An area of understanding may lack theory or some component(s) of theory. Furthermore, a theory or one of its components may (2) require conceptual refinement, (3) be rejected, or (4) be replaced. Finally, (5) the scope of theory may be expanded to include novel phenomena or to embrace formerly disparate phenomena. These five stimuli identify five kinds of questions that are *fundamental* to the establishment or advance of scientific understanding. The fifth area, expansion of domain or scope, also foreshadows the analysis of integration that we will explore in Chapter 7. We use the term "fundamental" here in the sense of a foundation for a scientific area, because understanding is built on the theory, the empirically observable phenomena, and the dialogue between them. The reason that we call these questions fundamental is because they arise from this logically derived list of possible changes in theory. *Fundamental questions are those that can lead to the establishment, refinement, rejection, replacement, or expansion in the scope of a theory or its components.* Within an area of scientific understanding, the most likely locations for action of fundamental questions are the conceptual constructs, or the "lines of text" in the dialogue between theory and observable phenomena (Fig. 6.1). Successful fundamental questions change the structure or content of understanding (Fig. 6.2). The focus of fundamental questions is within science itself. Other possible foci for questions, such as personal, societal, or political interest (Fig. 6.3), will be discussed later in the book. These other foci are legitimate and appropriate. However, we choose to discuss fundamental, science-focused questions first because they are central to advancing scientific understanding.

Because understanding requires theory as its touchstone, understanding is limited by the existence and status of theory. This limitation of understanding by theory is unavoidable. In spite of this necessary limit, there are many ways in which theory can help advance scientific understanding. Theory can be made explicit, scrutinized, and changed as a result of internal refinement and as a result of the dialogue between theory and observable phenomena. The possibility that understanding in an area is limited, incomplete, or incorrect means that a constant effort to identify and address fundamental questions is a crucial aspect of scientific progress.

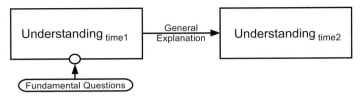

Figure 6.1 The relationship of fundamental questions to understanding. The fundamental questions relative to understanding at a certain time, t_1, generate a new understanding by changing the way that general explanation is cast to yield a new state of understanding at t_2. This is a general way to conceive of the interaction of fundamental questions with a system of conceptual constructs. A more specific suite of interactions between fundamental questions and understanding is show in Figure 6.2.

B. Basic Criteria for Fundamental Questions

In this section, we specify the criteria for fundamental questions in more detail. The ramifications of the criteria for fundamental questions must also be recognized. There is a practical reason for seeking criteria for fundamental questions. These criteria may help us to discriminate among the many competing questions ecologists pose. If there were some way to determine the significance of the many research questions competing for ecologists' time, the chances for advancing the discipline might be enhanced. Based on progress in other disciplines — for example, quantum physics, biology of DNA, and classical genetics — it appears that a degree of consensus on what questions are fundamental can lead to great progress (Darden 1991, Kuhn 1977, Ziman 1985). Of course, individual scientists or groups should feel free to investigate whatever questions they wish. We only point out that concerted effort focused on consensus-based questions can benefit a field as a whole.

We will suggest criteria for recognizing fundamental questions in ecology. If it is possible to establish criteria that have some objective basis, then this important task could be moved from the realm of opinion and belief to one that would be explicit but neither ad hoc nor

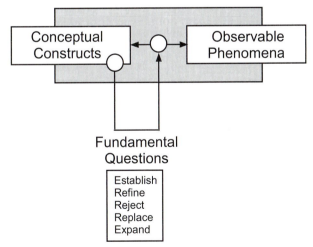

Figure 6.2 Fundamental questions can change understanding in one or more of the five ways in which they affect theory, including establishment of a theory *de novo*, refinement of an existing theory or key component of theory, rejection of a component or theory found to be defective, replacement of a faulty theory or component, and expansion of the domain of a theory.

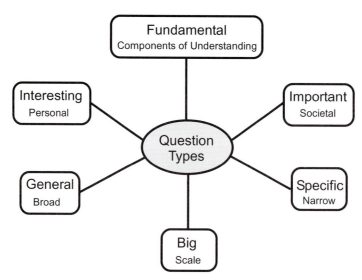

Figure 6.3   The types of questions. Fundamental questions relate to the components of understanding within an area of science. "Big" questions relate to a large spatial extent. The labels "general" or "specific" refer to the size of the conceptual scope or inclusiveness of the domain addressed. Important questions are those of societal interest, whereas interesting questions are motivated by personal fascination. These types of questions are not mutually exclusive.

authoritarian. It will not be possible to eliminate judgment from decisions about what constitutes a fundamental question, but it should be possible to clarify the basis for the judgments. We give some existing examples when introducing the characteristics of each stimulus for fundamental questions in subsequent sections, but we present examples of plausible contemporary fundamental questions in Part III. Every ecologist or group can, and should, nominate their own fundamental questions (Box 6.1), and the ones we explore here are only exemplary and not an exhaustive list.

1. Fundamental Questions Lead to the Establishment of a Theory or of a Missing Theoretical Component

When a theory is in a rudimentary stage, almost all of its components will have to be established (Loehle 1987b, Rosenberg 1985; Chapter 5). In some cases, the components will have to be generated *de novo*, while in other situations, they can be borrowed from other theories. This includes, of course, generating a firm empirical foundation in a given domain. The quantitative determination of patterns in nature is a necessary step in both new and continually developing areas of ecology (Peters 1986, Currie and Paquin 1987, Carpenter 1998). Later in the development of theory, only some components may be missing. In ecology, many of the existing theories are incomplete, perhaps because of the novelty of the science and the complexity of the subject matter. The theories often lack a broad scope (e.g., community theory; Brown 1981) and a clear statement of components (Rosenberg 1985), or clear connections between existing elements are missing (see DeAngelis and Waterhouse 1987 for equilibrium theory). Establishment of theory may, on one hand, require the identification of underlying assumptions (Murray 1986) or a clear definition of objects and relationships (Novak and Gowin 1984). Alternatively, as theories

---

BOX 6.1 Example of Questions for Advancing Understanding

We have abstracted these questions from the ecological literature but have not given authorship information because our purpose is to show the diversity of questions and the relative liberty ecologists take in proposing them. This liberty may not be entirely helpful because it invites propagation of questions at the expense of focused selection of those with greatest promise. Diversity is nevertheless a desirable state because it supplies material for discussion and selection. Focus can be achieved by relating questions to theory development or evaluation.

- A central task of community ecology is trying to understand which factors determine . . . richness.
- One of the key goals of ecology is to explain the distribution and abundance of species.
- A central issue of ecology is to what extent patterns can be extrapolated across scales.
- Large-scale synchrony in animal populations has become a key issue in ecology.
- A key question in ecology is which factors control species diversity in a community.
- A key issue in ecology is how patterns of species diversity differ as a function of scale.
- A central question of ecology is how community properties change . . . in space.
- A central aim of ecology is to explain the heterogeneous distribution of biodiversity on Earth.
- Identifying the factors controlling local community structure is a central problem in ecology.
- One of the fundamental questions of ecology is what controls biodiversity.
- Understanding how animals make decisions is a fundamental question in behavioral ecology, which has cascading effects on how animals respond to environmental variation.
- A fundamental question in plant ecology is if and how the intensity of competition changes with productivity.
- A fundamental question in ecology is how many species occur within a given area.

---

approach a more complete roster of components, they may require general models and unifying conceptual frameworks (Pickett and Kolasa 1989). Remarkably, we find clear and explicit frameworks to be rare in ecological theories (Schrader-Frechette 2001, but see Elser 2003 for a positive example). Cadenasso et al. (2003b) discussed the general nature of frameworks and applied the concept toward a theory for ecological boundaries.

The existence of a complete theory can be assessed both structurally and functionally. The fact that theories can be evaluated in terms of both their structure and their function provides another dimension for thinking about the need for the complete range of fundamental questions. Structurally, theory completeness can be assessed by comparison with Box 3.2, which lists the various possible components of theory. In well-developed theories, a large number of these components will exist. Even in well-developed theories, one or two components may be missing if their functional role is filled by other, existing components. For example, laws may be replaced by confirmed generalizations, as is the case in many ecological theories (Cooper 1998). Similarly, relationships can be replaced by models. Functionally, a theory can be judged via its explanatory, creative, or predictive success (Chapter 3). In other words, theory completeness must ultimately be judged by how well the theory works along these conceptual and empirical dimensions. If the complete roster of possible structural components is not present, a theory may still be successful. So the judgment of theory completeness is based first on functional completeness, which is served by the secondary criterion of structural or component completeness. Because success in using theories is predicated on their structural completeness and functional development (Loehle 1987b), this kind of fundamental question is critical.

2. Fundamental Questions Lead to the Refinement of Theoretical Components

As pointed out in Chapter 4, theory is not static (Loehle 1987b). All the components of theory, from the basic definitions to the high-level models and frameworks, are subject to improvement. Conceptual clarification provides improved ways to think about the basic ideas of a theory, improved ways to articulate them, or better statements to define the conventions. For example, conceptual clarification was critical in advancing genetic theory. Early genetic research was based on the relatively loose concept of unit characters, which evolved to the idea of genes. This was a key step in the development of the science of genetics (Darden 1991).

A common problem in ecology is lack of clear definition of the objects addressed by a theory (Brown 1981, Cooper 1926, Loehle 1983, Pickett et al. 1987, Reiners 1986). Sagoff (2003) was explicit in revealing this problem in his critique within the context of an ecosystem: "no theory can be tested unless it defines the class of objects the behavior of which it seeks to understand." For example, the question, "Do ecosystems have a characteristic trophic structure?" cannot lead to a useful answer if an ecosystem concept includes a cup of yogurt, a pile of dung, and a great lake, and no framework is available for sorting out such divergent kinds of structures. For the moment, however, we will leave aside the complex issues of ecosystem conceptualization and focus on an example of the study of disturbance that we deem more tractable.

The initial conceptualizations of disturbance were intuitive and linked to the community realm of ecology. However, disturbance theory currently demands broader and clearer definitions of the phenomenon to permit cross-system comparisons and predictions (Dale et al. 1999). Disturbance in its original community context was seen as death of dominant individuals (Sousa 1984). Contemporary disturbance theory defines the process as an event that disrupts the structure of an ecological system. Of course, this definition requires that a clear articulation of the specific system of interest be made. In other words, a model of the system structure must be specified. When such clear generalizable definitions (Pickett and White 1985, Pickett et al. 1989, Rykiel 1985) are widely accepted, the understanding of disturbance can advance because more instances can be successfully accommodated by the theory.

Attention to refined definitions can have other substantive benefits in ecology. For example, the need to clarify troublesome concepts in community ecology led to the invention of the ecosystem concept by Tansley (1935) and a subsequent burgeoning of a new approach to ecology (Golley 1993, Likens 1992, Odum 1971). Considering how the ecosystem concept came to be used by many ecologists further exemplifies problems with definitions in ecology. Some ecologists have defined the ecosystem in a way that assumes certain characteristics. For example, ecosystems have been construed as materially closed, or self-regulating and homeostatic, or autotrophic. The core definition implies no such assumptions. If some particular application of the concept justifies assumptions such as these, those assumptions should be a part of the model that applies the general, abstract definition to specific cases or situations (Pickett and Cadenasso 2002).

We propose that clarity in definitions of the core or foundational concepts in ecology, such as ecosystem, community, and the like, can be advanced by making those definitions as general and free of assumptions as possible. Of course, assumptions will be a necessary part of applying the definitions to specific computer models, experiments, or field observations. Furthermore, the need for making general definitions should not become synonymous with making them all inclusive because such definitions invite confusion. For example, a definition of ecosystem as an entity where an entity is a group of subentities linked by a process (Scheiner et al. 1993) is a step in the right direction. However, since this definition does not clarify what "linked" means, anything could be an ecosystem. But "anything" cannot have common, identifiable ecological properties that would add to the development of theory and thus understanding. More work is needed on this general definition, and work by Jax (1998) could assist with this task.

Jax (1998) used the idea from philosophy of science that *definition* is the process of stating an abstract concept applicable to many scales and situations, while *specification* is the act of limiting a concept to one case or class of cases. We now recognize that ecological concepts — or perhaps it is better to say the terms that stand for the concepts — have at least two dimensions: meaning (the core definition) and model (Pickett and Cadenasso 2002). Clarity in definition is complicated by a third aspect or connotation of ecological terms. Not only do the most important and widely applicable terms from ecology have their core definitions and their model specifications, they also have a metaphorical dimension. When ecologists communicate outside their discipline, say to the public or even to other scientists unfamiliar with ecological concepts, metaphor comes into play. Metaphor is, of course, a figure of speech that takes one thing as a different, dissimilar thing.

Metaphorical uses of the term "ecosystem" include simply standing for a place, or for a uniform patch of habitat, or a system with closed nutrient cycling loops, or a homeostatic, persistent system. In the popular mind, the term "ecosystem" is often used to connote balance or persistence. Some uses of the ecosystem explicitly exclude places where people are present. This is but a small sampling of the many metaphorical uses of the term "ecosystem." Some of these metaphorical uses help in engaging the public in dialogue or in opening communication with other disciplines. However, these and many other metaphors can hide ideas that are really assumptions that shape the model specifications of the ecosystem. They may not turn out to be correct or even applicable to the subject of a public discourse about ecology.

It is important to know about this metaphorical dimension of ecological concepts so it can be used effectively to promote public dialogue and interdisciplinary communication (Pickett et al. 1999). But it is also important to know that such metaphorical uses often embody assumptions that are inappropriate or incorrect. The subsequent dialogue has to sort out these complexities of different assumptions that various participants in a conversation may have. It is such metaphorical uses, with their hidden assumptions, that are one reason that Bohm (1996) called for broad scientific dialogues to seek common meaning rather than merely common language. Common language often operates at the level of metaphor and obscures differences between meaning and model.

Another ecological term that is being used metaphorically to promote interdisciplinary dialogue is "patch dynamics." The metaphorical use of "patch dynamics" illustrates another caution that must be kept in mind when scientific terms are used metaphorically. The term "patch dynamics" is being used to broaden the communication between urban designers and ecologists (McGrath et al. 2007). One reason that it is useful in linking design fields, such as landscape architecture and ecology, is that designs focus on discrete parcels or projects. Patch dynamics helps ecologists, who study the spatial structure and dynamics of systems, to communicate with professionals who design and build the spatial structure of human settlements. However, the term has metaphorical limits. Contemporary thinking about patch dynamics in ecology (Pickett et al. 2000) recognizes that spatial heterogeneity is often continuous and can be represented and modeled as fields and gradually undulating surfaces whose elevations represent differences among locations rather than as discrete patches. In other words, the term has grown in ecology to include not only the discrete forms of heterogeneity it was born to serve, but gradual forms of heterogeneity as well. Sophisticated dialogue both within ecology and between ecology and other disciplines must recognize the new technical implication of the term "patch dynamics."

The recognition that ecological terms have a core meaning, a range of model specifications, and a metaphorical dimension helps with the job of refinement in ecological theories. We now move to the kind of fundamental question that deals with replacement of problematic components of theories rather than the construction of new components or refinement of existing components that we discussed earlier.

3. Fundamental Questions May Lead to the Rejection of Inadequate Theories or Inadequate Components of Theory

Questions that successfully challenge the completeness, applicability, or coherence of a theory can have a profound effect on understanding. The observation that communities often intergrade rather than change abruptly contradicted one of the tenets of early plant community ecology theory. Such a contradiction implied the fundamental questions: "Are communities only rarely discrete entities?" or "Do communities exist?" (e.g., Palmer and White 1994). Similarly, the demand by the classical theory of succession that the process be driven by its endpoint (Clements 1916) inhibited discovery of successional mechanisms, such as the role of herbivory and predation, which is now considered to be important and broadly applicable. The work of some of Clements's (1916) ecological contemporaries (e.g., Cooper 1926) dealt with an implied underlying question: "Does climax drive succession?" This is a fundamental question that stimulated research and conceptualization and that broke the hold of a dominant research paradigm.

A well-developed theory or one of its components that does not correlate adequately with reality and with other aspects of the theory is an impediment to understanding. However, if well used, incorrect theories or components of theory can have great heuristic or creative value (Weissman 1989). The data generated to address them, the conceptual refinements they generate, and, indeed, the useful components that survive the demise of the incorrect theory can be of use.

We present some examples of the value of flawed theoretical structures. Incorrect theories or components are often the stimuli to develop correct and useful theories. For example, the original theory of continental drift proposed by Wegener was incorrect in the majority of its details; however, its broad outline was the basis of the currently accepted theory of plate tectonics (Cohen 1985). Components may also be rejected because they are found to be inappropriate or irrelevant as a theory develops. An example comes from genetics. In classical genetics, the early emphasis was with dominance and recessiveness. There were several features that led to its rejection: (1) the generalization was not found to be universal, (2) there was no mechanism in the theory for it, (3) its absence did not alter development of the rest of the theory, and (4) its potential explanation was speculated to be on a level of organization outside the domain of the theory (Darden 1991). Contemporary genetics has a more subtle view of the relationship of alleles. This example brings us to the idea that theoretical components found to be faulty or lacking are often replaced.

4. Fundamental Questions May Lead to the Replacement of a Theory or Its Components

The rejection of inadequate theoretical constructs, while removing an impediment to understanding, may precipitate a crisis (Kuhn 1970). Such crises indicate a limit to understanding. Only with the replacement of the erroneous or inadequate theory or component will understanding again grow (Cohen 1985). Thus, fundamental questions that lead to replacement of the problematical aspect of theory result in a clear advance. Replacement is not a simple act. It requires, of course, that a clear and relevant alternative be available. The demise of the erosion cycle in geomorphology provides an example. This cycle, proposed by W. M. Davis in 1899 (see Chorley and Haggett 1965), emphasized progressive development of landforms from youth to maturity to old age. Young landforms were steep and highly topographically divided. Tectonic uplift was fast, and so erosion was rapid and deep. In contrast, mature landforms were rolling or nearly flat, with low rates of erosion and no tectonic uplift. The replacement of this theory was incomplete until the elements of the new theory of process geomorphology, focusing on the equilibrium between deposition and erosion and the factors that affect it, became well developed and articulated in the 1960s (Gilbert 1980, Goodlett 1969, Hack 1960). Note that Davis's original theory was phenomenological, and its replacement was mechanistic.

Likewise, the inadequacy of the Lotka-Volterra models as cores for community theory begs their replacement. The attempts by Ginzburg and Akçakaya (1992) to establish alternative equations based on the assumption that predation involves an interaction among predators is an example of a possible radical replacement. Hubbell's (2001) rejection of the empirical encumbrance that comes with the consideration of differences among species and their replacement with an assumption of equivalency among species is another such attempt. Although its prospects and ultimate impact are yet unknown, there is little doubt that Hubbell's approach represents a radical departure from the traditional efforts in community ecology theory (Norris 2003).

5. Fundamental Questions Lead to Increasing Scope of Theory

A question that causes a theory to encompass some phenomenon well outside its accepted domain is fundamental. Such unification may absorb another theory or embrace a phenomenon that did not possess its own well-developed theory. For example, plant ecologists came to realize that those changes in plant communities that were not noticeably directional resulted from the same basic processes that produce directional succession; this was a significant broadening of succession theory (Miles 1979). Note that such significantly expanded theories may retain their narrow names, leading the uninitiated to continue to assume the now superceded scope. Succession theory these days is equivalent to vegetation or community dynamics theory. In fact, community assembly adopts essentially the same kinds of processes that succession theory uses.

An example of extension of scope also appears in other theories. The possibility that disturbance theory can be cast in the same terms as predator/prey theory is a significant expansion of both those areas (Petraitis et al. 1989). Island biogeography theory has also been expanded. Extension of the domain of island biogeography to include other kinds of isolated, island-like habitats such as lakes, mountaintops, or forest fragments (Brown 1971, Forman et al. 1976, Keddy 1976, Quinn and Harrison 1988) is another example of how a fundamental question may influence understanding in ecology. In this case, the question was, "Does the model of MacArthur and Wilson apply to situations other than oceanic islands?" or "What components of the original island biogeographic model can be generalized to other kinds of habitats?" The enriched theory shows new potential for expansion and interaction with patch dynamic models (Rogers 1997).

An example of how the state of a theory was changed by recognizing the need for new components appears in community theory. The Lotka-Volterra models require a fine-tuning of parameters to obtain coexistence of two or more competing species. Such fine-tuning is unrealistic for natural systems because parameters are likely to fluctuate in space and time. Note, however, that the Lotka-Volterra models originally assumed, implicitly, that patterns and interactions were homogeneous. By extending the domain to include regional dynamics and adding corresponding theoretical components, Caswell (1978) managed to produce robust patterns of coexistence. The fundamental question implied by this accomplishment is, "What must be added to the Lotka-Volterra models to make them better reflect natural processes?" We suggest that these sorts of questions will apply to many other ecological domains. The expansion of the scope of a theory, if that expansion involves all or part of some very different theory, such as those mentioned previously, is a case of integration. We will expand this concept in Chapter 7.

The criteria just presented constitute all logical stances toward change in theory. In practice, the approaches to altering theory may be applied sequentially or simultaneously. Thus, it may be difficult, and indeed unproductive, to attempt to isolate the different kinds of change in analyzing past progress or the potential for future improvement of understanding. Because this roster of changes is logically complete and influences theory as a pillar of understanding, the questions that motivate them are truly fundamental to the progress of science.

III. Examples of Fundamental Questions

The discussion so far shows that to evaluate the fundamental questions in a discipline first requires an analysis of its theory. Specifically, how complete and well developed the theory is (cf. Chapter 5) determines what the fundamental questions will be. We briefly present several cases to show how ecologists might determine fundamental questions. The examples are (1) island biogeography, (2) ecosystem energetics, and (3) disturbance theory. Furthermore, we suggest a fundamental question about each of these areas that might exemplify the role of fundamental questions in unifying domains. We cannot present these theories in all their richness here. Details of the individual theories can be found in general texts or in the specialized references we cite for each one.

A. Island Biogeography Theory

This theory is well developed, with a quantitative core (Abbott 1980, Brown 1981, MacArthur and Wilson 1967, Simberloff 1974, Wu and Vankat 1995). Its phenomenological predictions have been tested numerous times, either by experiment or by comparison with island patterns in nature (Simberloff 1974, see also Castle 2001 for an extensive analysis of the status of island biogeography theory in ecology). The search for generalizations or factors that can complement or explain patterns predicted by the original equilibrium models is a fundamental focus suggested by the theory. We list, in a general form, several fundamental questions that address problems in the theory. The originating fundamental question dealt with how island characteristics relate to species diversity.

1. How unlike oceanic islands can terrestrial patches be and still be included in the scope of the theory? This question deals with the assumption of the theory that islands are discrete and surrounded by a completely inhospitable medium. If a gradient of isolation is assumed, a related fundamental question is suggested, as follows. If the surrounding medium is only partially inhospitable or becomes gradually more inhospitable with distance, will the theory still hold?

2. Is the current formulation of the theory a subset of a possibly broader theory accounting for patterns of species richness along a general axis of increasing habitat isolation? This question is an extension of number 1.

3. What is the balance among extinction, colonization, history, and habitat heterogeneity as causes of species diversity on islands and island-like habitats? This question examines the assumption that islands are uniform, that all dynamics are based on current invasion and extinction, and that the conditions on islands do not change sufficiently over time to affect the invasion and extinction rates.

4. Are there other factors that must be incorporated into the models to apply the theory more closely to specific cases or to new sorts of patches and matrices between patches? This question opens the way to a larger theory that can address heterogeneous patchworks with neighborhoods of varying degrees of hospitality to the organism or process of interest. Such a theory might be a better foundation for conservation biology than standard island biogeography theory (Gilbert 1980, cf. Soule and Kohm 1989). An additional feature of such a broader theory is that the gradients or boundaries between habitat patches may be important functional components of the systems. An emerging boundary theory (Cadenasso et al. 2003a, 2003b, Strayer et al. 2003) may complement or be subsumed by the broadened theory of island biogeography. Finally, scale modulation (Kolasa and Waltho 1998) of the differential perception of patch quality or inhospitability by different species may be required to complete the necessary set of dimensions such a comprehensive theory would have to cover.

5. At which spatial scale(s) do the assumptions, mechanisms, and predictions of the theory fail to apply? This question suggests that the distances of dispersal dynamics relative to the spatial structure of a patchwork may become an important part of the theory suitable to addressing diversity in landscape patchworks. The application to patch coral reefs (Waltho and Kolasa 1998) is a case in point that illustrates this fundamental question.

The fundamental questions about island biogeography theory reflect two difficulties with the original theory. First, the new fundamental questions address contradictions discovered in testing the theory. The new set of fundamental questions also emerges from a broadening of the scope of the theory. Problems of application to island analogs such as continental patches, insects on individual plants, and ephemeral patches have also generated new fundamental questions. A critical concern of these fundamental questions is the nature and impact of mechanistic constraints, such as history and contrasting species behaviors, on predictions of species richness on islands. At the very least, these fundamental questions point out that, in applying island biogeography theory, it is important *not* to use the equilibrium core without translation to the contingencies of specialized cases.

B. Theory of Ecosystem Energetics

Attempts to find a general, unifying explanation of the organization of ecosystem structure, function, and change through time have a long tradition in ecology (e.g., Lindeman 1942, Lotka 1922, Odum 1953). Energy transfers and transformations are clearly fundamental to all biological and nonbiological processes. Consequently, focusing on energy as a central currency, with thermodynamic laws as operating principles, seemed a logical step in the development of a theory of ecosystem organization. The thermodynamic conjectures of H. T. Odum (1983) have played a central role in ecosystem theory. In particular, energy is seen as a universal explanatory currency, and the maximum power principle is seen as a universal explanatory principle. The maximum power principle states that systems develop so that power is maximized. Power is equated with energy flow.

The two basic assumptions of Odum's theory have been criticized (Mansson and McGlade 1993) for two reasons. First, obtaining operational relationships between abstract, thermodynamically governed principles and real energy and material flows, with their myriad forms, uses, aggregations, and transformations, has been particularly problematical. This difficulty calls into question the validity of using energy as a single universal explanatory currency. Second, it seems that empirical data on energy flow in real ecosystems, as well as in dynamic simulations of ecosystem behavior, contradict the maximum power principle.

The problems with the maximum power principle raise fundamental questions about the role of energetics in ecosystem theories:

1. How should we interrelate the energetics of organisms to the thermodynamics of entire ecosystems (Nisbet et al. 2000)?

2. How do we deal with the diversity of forms of energy, and their interconversions, if they are not reducible to a singular form?

3. Do only some parts of systems conform to the maximum power principle, whereas others do not?

4. In a broader context, how do we interrelate energy flow with material fluxes (Reiners 1986) and changes in abundance with fluxes of organisms in ecosystems (Elser 2003, Sterner 1995)?

5. How can we relate energetic theories to theories of ecosystem organization based on such principles as entropy (Brooks and Wiley 1988) or information flow (Ulanowicz 1986, 1997)?

6. Do ecosystem energetic regimes constrain evolutionary processes and vice versa?

7. Is ecosystem development (sensu E. P. Odum) caused by energetic processes or are energy flows determined by community membership and abiotic transformations?

8. What are the principles by which energy units are aggregated into operational compartments?

Aggregations by taxon, trophic position, or functional group all have major problems and inconsistencies which, should they remain unresolved, will continue to hinder the study of ecosystems.

C. Disturbance Theory

Although some ecologists have recognized physical disturbance and other disequilibrating factors for a long time (e.g., Andrewartha and Birch 1954, Cooper 1913, Watt 1947), such factors became the subject of concerted theoretical development much more recently (Bormann and Likens 1979, Choi et al. 1999, Pickett and White 1985, Rykiel 1985, Sousa 1984a, 1984b, Walker 1999). Thus, many of the components of disturbance theory remain tentative (see Peet et al. 1983). There is a need to clarify the basic concept of disturbance and to develop a way to translate the concept to a variety of spatial scales and organizational levels (Rykiel 1985). This is especially true based on the expansion of the concept beyond its original scope (Pickett et al. 1989). Some of the fundamental questions that emerge are these:

1. How does disturbance affect system attributes other than species richness? Nonspecies parameters include such variables as system architecture, trophic and nutrient dynamics, and erodability. The vast majority of the generalizations about disturbance focuses on species richness as the response variable of interest, but disturbance resets many portions of the ecosystem to earlier successional stages and thus entails a host of concomitant changes.

2. Is the intermediate disturbance hypothesis a suitable organizing idea for all system attributes? The intermediate disturbance hypothesis has been a central idea in the theory of disturbance. It proposes that species richness is maximized at intermediate frequencies or intensities of disturbance (Connell 1978, Huston 1979). Part of the problem with applicability of the hypothesis is translating it into operational terms. Intermediacy begs the question of specifying upper

and lower limits of disturbance frequency or intensity. Assuming it is possible to erect reasonable translation(s) for richness, is the translation the same for all variables of interest?

3. What factors might prevent attaining maximum species richness at intermediate disturbance intensity? The examination of the intensity and the effects of disturbance along environmental gradients are an important approach to answering this question.

4. How do disturbances and other system-organizing disequilibrating factors interact? Here, the common effects of a range of apparently contrasting processes needs to be examined. For example, since disturbance focuses directly on structure, one might ask, how does stress — a focus on function — act as a disequilibrating factor? Is its effect similar to disturbance? Indeed, when and how does it act *through* disturbance?

5. Can disturbance be analyzed in the same terms for systems of differing levels of organization? This question reflects ongoing interest in how to apply the idea of disturbance outside the community realm in which it was first developed. For instance, how does the disturbance concept apply to landscapes, ecosystems, populations, and individuals (Kolasa and Pickett 1989, Pickett et al. 1989)? In particular, there is a need to conceptualize disturbance in human-dominated ecosystems where it may not be productive to consider people themselves to be a "disturbance."

6. Are disturbance and variation of system components different expressions of the same underlying process? Can the processes be unified in a single theory?

The illustrative fundamental questions we have listed for disturbance theory reflect the developmental state of this theory and the need for both extensive empirical study and conceptual refinement. Because disturbance theory is clearly closely associated with spatial processes, we can pose a bonus question: "Are patch dynamics, metacommunity theory, and island biogeography spatial scale-dependent varieties of the same higher level theory of organisms moving in space (Fig. 6.4, Box 6.2)?"

IV. All Fundamental Questions Are Not Created Equal

The previous sections presented several criteria for establishing fundamental questions that have to do with the nature and status of theory in an area. However, other aspects of fundamental questions help us judge their suitability for research, and they are discussed in this section.

Recognizing a question as fundamental because it fulfills one or several of the logical approaches to theory is a necessary but insufficient condition for adopting the question as a high priority in a domain. For a question to be given high priority, it must also (1) be based on sound and clear concepts (Novak and Gowin 1984) and (2) have the capacity to advance generality by

---

BOX 6.2 Three Separate Theories Differ Essentially Only in Scale along Three Dimensions

We suggest that all of these theories and perspectives could be reformulated as *a function of habitat resolution scale, time, and barriers to organism dispersal* and settlement at a given scale (Fig. 6.3). By modifying parameter ranges of these three dimensions, one should be able move along among their original scopes. What these parameters should be remains a challenging technical issue. A nice by-product of this prospective unification would be the emergence of a theory capable of dealing with a continuum of isolation, time, and habitat resolution as opposed to the current conceptual fragmentation.

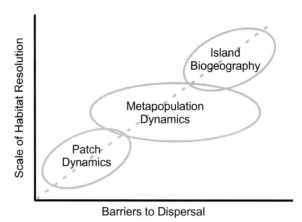

Figure 6.4   Patch dynamics, metapopulation dynamics, and island biogeography theory differ primarily by scale of habitat resolution and the magnitude of barriers to dispersal of organisms. These differences have traditionally defined the expectations about patterns and influenced the focus on the mechanisms involved even though the mechanisms and patterns change gradually along the same gradient. The time axis reflects investigator association with these theories more than any fundamental difference as all categories of phenomena can occur simultaneously.

encompassing a broad scope (Kuhn 1977, Mayr 1982, Slobodkin 1985). A question may indeed be fundamental, but it may be so poorly phrased or broadly couched conceptually that it impedes understanding. A question about the relationship between vague objects or poorly articulated phenomena may create confusion, mislead the unwary, and be difficult to test. Questions about the relationship between diversity and stability, although potentially fundamental, ultimately were abandoned by most ecologists for three decades because of the lack of clear conceptual bases and definitions for diversity and stability (Brown 1981). Only more recently, as a result of conceptual trimming by restricting stability to temporal variability, research into diversity-stability hypothesis has seen some progress (e.g., Tilman 1999). Questions about intermediate intensities of disturbance are in danger of the same fate (Pickett and White 1985). Both these difficult situations suggest that fundamental questions in these two areas needing immediate attention would be ones that better establish the basic concepts, definitions, and interactions needed for a theory.

These examples show how an awareness of the status of a theory, especially if it is a nascent one, can assist in ranking questions for attack. The question about the relationship between diversity and stability may indeed be fundamental, but still more fundamental (in the literal sense of a foundation) is the question about the nature of the objects of study and the relationships that might connect them (Ulanowicz 1986). Unresolved questions about the nature of objects in ecology (Box 6.3) therefore require urgent attention (Kolasa and Pickett 1991, Reiners 1986) before they result in other fiascos like that committed with diversity and stability, in which functional diversity was erroneously replaced with structural diversity.

The need for conceptual clarity is especially crucial when a question originates in a societal concern. For example, it is appropriate to use metaphorical terms in the public discourse, and the public has stated an interest in such metaphorical terms as "ecosystem health" or "ecosystem services." Such terms can be given technical definitions, but it is important to realize that such specifications often develop long after the social conversation about these issues has begun. It is also important to recognize the complex social values that are implied in terms such as

---

BOX 6.3 More on Ecological Objects

We expand our comments on the nature of ecological objects because their sound conceptualization is of crucial importance to the development of the discipline, while inadequate conceptualization is likely to lead to unproductive and wasteful labors.

The nature of objects in ecology has been a vexing problem almost from the beginning of systematic ecological inquiry. Clements (1916) and Gleason (1917) are widely remembered for their contrasting views on the nature of multispecies systems, specifically on whether they should be approached as organized entities or contingent collections of interacting components. Despite decades of considered deliberations, pattern analysis, and development of relevant theories (succession, assembly, ecosystem), the answer to this question is far from definitive. Some progress has been made and resulted in the recognition that the organization of communities as self-identifiable entities is a matter of degree (Grimm 1998, Jax 1998, Scheiner et al. 1993). However, whether multispecies assemblages form integrated entities, albeit weakly, or just haphazard collections of independent components decides many methodological issues. Decisions on what to measure, how complexity and stability may relate, what is the appropriate scale for assessing mutual impacts of one "thing" on another, how to treat dispersal into a community, what is the best framework for representing material cycles, how communities come into being, and many others all depend on a priori identification of the objects. For example, it would serve little theoretical purpose to analyze coordination of components (e.g., synchrony of populations, Red Queen hypothesis of the coevolutionary ratchet) or system stability if species were not part of a system (cf. Mahner and Bunge 1997) or if only some were arbitrarily included but not others. Thus, it is reasonable to conclude that the answer to this general question will have a profound and broad impact on how these related theories develop, what other questions must be asked or ignored, and how ecology delivers its findings to the society at large.

"ecosystem health" and to go beyond the metaphors in framing fundamental questions for scientific research. We will return to the public dialogue in a later chapter.

Fundamental questions may also be ranked by how much impact they would have on an area of understanding. If answering a fundamental question would fill a greater gap in the body of conceptual constructs or would generate a cascade of creative work on a problem, it should be given higher priority than other questions. In addition, fundamental questions that are more feasible to answer can be given higher priority. In the case of infeasibility of a truly fundamental question, it may be wise to break the question down into more feasible components.

Given the desirability of generality in science (Jacob 1982, Mayr 1982, Murray 1986), a question that has the potential for unifying apparently disparate areas is more fundamental than a question that keeps to a narrower scope. However, for such questions to be successful, the theoretical and empirical foundation in the areas to be unified may already have to be well developed. Alternatively, the fundamental question suggesting unification may lead to a greater empirical and theoretical development in the areas to be unified. Thus, fundamental questions may be significant to the advance of a discipline even when they are far from being answered. The nature of unification and integration across domains will be developed in Chapter 7.

A fundamental question may lead a discipline down an incorrect track or toward the elaboration of an incorrect theory. Such error may not be a great disaster, however. Historians and philosophers of science agree, despite their divergent assumptions and approaches, that science is a self-correcting process (Kuhn 1970, 1977, Longino 1990, Popper 1965). If a fundamental question leads to a well-developed, complete theory that is wrong, the completeness and high degree of development of the theory can lead to the very clarity, hypotheses, and tests that may condemn it (Lewis 1982, Loehle 1987b). The potential for replacement of erroneous theory is part of any ideal scenario of scientific progress.

To this point, this chapter has presented the five criteria for recognizing fundamental questions. The five criteria have involved the status of theory, since they lead to establishment, rejection, replacement of theory, or increasing its scope. The fulfillment of at least one of those criteria is necessary for a question to be considered fundamental. However, those criteria are not sufficient for a question to be embraced as a guide for action. Additional refinements must be made if a question is to be recognized as fundamental and is to be acted on.

This section has identified fundamental questions as those incorporating (1) clear sound concepts and (2) broad scope. Fundamental questions that deal with scope may suggest a linkage with other areas within ecology.

V. Where Do Radically New Theories Come From?

Because we have accepted a broad, contemporary conception of theory, we hope that the fundamental questions suggested by establishment, rejection, refinement, replacement, and expanding scope of theory can encompass all the major concerns of a discipline. However, we must recognize that sometimes the most significant questions in a science, those that lead to the establishment of an entirely new theory, will arise from outside a recognized theory. Perhaps the most revolutionary of such questions arise from outside any theory at all or even outside a nascent theoretical framework. Such cases are extremely difficult for practicing scientists to analyze and are likely to be impossible for anyone to predict. Our analysis of fundamental questions has focused on those cases in which the existence or emergence of a theory suggests the fundamental questions in that area. An example of a fundamental question from outside any theory might be one in physics that asked, "Can electricity and magnetism be unified?" This seems to have been the question that motivated Maxwell to conduct his famous experiment. Although his concern with this unification may suggest the glimmer of a theory, his insight was completely novel and was not deduced from any existing theory (Cohen 1985). Such insights must be considered inductive, since they do not arise from even a preliminary theoretical formalism, although they are founded on certain observations and, perhaps, on pretheoretic notions. We must therefore admit that the framework advanced here for fundamental questions does not cover the inductive stroke of genius in creating a theory or recognizing a novel domain.

Another kind of genius stroke involves taking an insight from an entirely unexpected source and applying it to generate new insight in an area. The term "abduction" introduced (with tongue in cheek?) by Hanson (1961) is more appropriate for such strokes of genius. The scenario involves (1) encountering some surprising results or phenomena; (2) noticing that a hypothesis (or perhaps a theory) of a certain type would alleviate the surprise and, in fact, encompass the otherwise surprising phenomena; and (3) developing a cogent hypothesis (or theory) of the noticed type to embrace the phenomenon.

In ecology, fundamental questions emerging outside either established or nascent theories that lead to the establishment of a theory may appear as musings about pattern. Peter W. Price (personal communication) proposed that fundamental questions were those that, in fact, led to the establishment of unprecedented new theories. Such questions have the general form, "What are the patterns of ecological phenomenon x, and what mechanisms generate those patterns?" Even in such a case, the criterion of breadth of scope should be applied. All areas of ecology can be encompassed within such broad general questions. The pattern/process form of such questions is analogous to a pattern/process definition of ecology. For example, "The central goal of ecology is to understand the causes of the patterns we observe in the natural world" (Tilman 1988). Of course the phrase "natural world" in reference to ecology implies organisms and the systems containing them. Based on Price's view, the bulk of our analysis of fundamental questions is relevant to the critique of existing theories and the competition among alternative theories. Coming up with a completely new pattern/process question, or encompassing disparate areas within a single pattern/process question, occurs in the same unmappable realm of inductive or abductive (sensu Hanson 1961) genius that all sciences must cherish.

Can ecology foster the inductive stroke? James H. Brown (personal communication) has suggested that certain characteristics fit such strokes in ecology. Using Robert MacArthur as an example, Brown identified four such characteristics: (1) attention to and discovery of general patterns, (2) ability to abstract the common elements and their essential relationships, (3) willingness to question accepted generalizations and extant theories, and (4) a sense of how the system works, derived from a familiarity with its natural history. These talents are closely related to several critical aspects of theory we have identified as advancing scientific understanding in general (Chapter 2): (1) generalization as both unification of disparate phenomena and summarization of a body of accepted fact, (2) idealization as the isolation of the phenomenon of interest from confounding influences in nature, and (3) abstraction of the essential elements and relationships in a system. The sense of how the system works is derived in part from the wisdom originating in natural history, but can be expressed in (4) the hierarchical structure of theories relevant to complex systems. Such a structure can accommodate the raw patterns captured directly from nature, and the abstract insights derived from them, that are the conceptual essence of theory. General theories contain more specific models that account for boundary and initial conditions relevant to particular cases (Chapter 3; e.g., Pickett et al. 1987, Pickett and McDonnell 1989). The message here is that truly revolutionary questions may arise from outside existing theories, and creative thought and research outside recognized theoretical foci may be immensely productive. Therefore, our analysis should not be misconstrued to suggest squelching such work.

VI. Conclusions and Prospects

The changes in understanding that have been outlined in this chapter are those that occur within a single area of understanding. Such areas are encompassed by a domain as described in Chapter 3 on the anatomy of theory. Understanding can change as a result of changes within a theory and of changes required by the dialogue of theory with observable phenomena. This dialogue often leads to articulation of fundamental questions. Fundamental questions are those that lead to the establishment, refinement, rejection, replacement, or extension of a component of theory or of an entire theory. Successful fundamental questions must, furthermore, employ clear well-defined concepts and address a broad scope. Clarity of concepts requires us to recognize that the terms scientists use to label concepts will have three dimensions: the core meaning, the reference to specific models or applications, and metaphors from which they emerge or which they suggest. Specific and explicit statements of the theory and identification of its components and their connections are required to identify fundamental questions. We have not considered radical unifications that embrace very different areas within ecology, yet ecology is a diverse discipline. Indeed, the potential for integration across radically different areas within our discipline is one that needs to be exploited. This is the topic of the next two chapters.

7

Integration and Synthesis

*"Science is a spider's web. Confidence in any one strand of the web
is maintained by the tension and resiliency of the entire web."*
Raymo 1991:144

I. Overview

We have shown how understanding depends on the changing characteristics of the theory within a domain. In the previous chapter, we indicated that there are five specific kinds of change possible in a theory. Fundamental questions are those that focus on one or more of these five kinds of change and so lead to a change in understanding. In this chapter, we move beyond the domain of a single theory. Here we examine how understanding changes through interaction among *different* domains. We also identify the kinds of questions that can drive changes in understanding through integration of theoretical frameworks.

What is integration? We have noted that ecology embraces diverse domains of understanding. Integration results from the combination of two or more different areas of understanding or their components into a new understanding. This chapter lays out the kinds of integration within ecology and identifies the kinds of fundamental questions that motivate integration.

Effective integration in ecology may require that disparate paradigms within the discipline be brought together. To analyze such an option of radical integration, we will present the meaning of the term "paradigm" and will identify the disciplinary paradigms that are contained within the science of ecology. Specifying the disparate paradigms helps articulate an ideal toward which integration can aspire. We identify possible fundamental questions to guide cross-paradigm integration and indicate the analogy with those fundamental questions that guide the growth of understanding within a subject area in ecology.

II. Integration

In developing our argument about integration, we have dealt primarily with changes within one area of understanding. However, ecology encompasses many subject areas, each characterized by a particular slice of ecological understanding (Fig. 1.1). Because the potential for integration across this broad array of subjects is so great and because integration and synthesis are so highly prized and have been so powerful in the history of science, explicit consideration of this problem

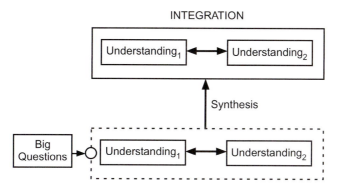

Figure 7.1   Additive integration. Integration can result from combining two complete areas of understanding by linking their complete theories. Such additive integration is said to result from posing a big or general question, which refers to a large conceptual extent or to a general phenomenon, respectively. The process leading to the state of integration can be labeled "synthesis."

in ecology is worthwhile. The dynamics of theory change involving two or more areas of understanding is integration. *Specifically, integration is the explicit joining of two or more areas of understanding into a single conceptual-empirical structure.* It logically follows that such integration can also result from an improvement in an already existing connection and the ensuing refinement of the conceptual-empirical structure. Changes involved in integration are analogous to the changes *within* an area of understanding, which we examined in Chapter 6. Like understanding within an area, integration across areas is a state established by a process, which can be labeled synthesis.

A. Modes of Synthesis

Integration arises in two ways. First, two areas of understanding may be combined more or less intact into a new, composite understanding. Such integration can be called *additive integration* (Fig. 7.1). Additive integration at the level of models is illustrated by the combination of autecological information about plants with forest stand dynamics to generate individual-based simulation models of communities. Theories of induced defenses in sessile animals and plants might be combined into a unified area with specific translation modes for the idiosyncrasies of each group. Additive integration is also illustrated by the developing theory of ecosystem engineering (Jones et al. 1994, 1997), which combines an understanding of how organisms change the abiotic environment (via, for example, the formation of physical structures that act on abiotic flows of energy and materials) with the rich body of extant understanding of how the abiotic environment affects organisms.

The second approach to integration is a selective use of knowledge. Two or more areas of understanding may provide components that are combined to yield a new understanding. Such integration can be called *extractive integration*, because only certain elements of other areas of understanding are extracted and combined into a new state of knowledge (Fig. 7.2). Such integration and combination may be gradual or rapid. An example of extractive integration takes key parameters controlling plant-insect herbivore interactions and key parameters controlling plant-microbial pathogen relationships and combines them to produce a new, unified understanding. The growing interest in organismal stoichiometry as a factor in ecosystem ecology (Sterner and Elser 2002) is another example of extractive integration. This integration takes the

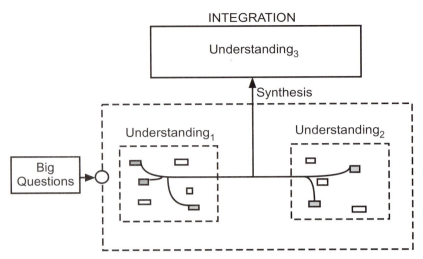

Figure 7.2   Extractive integration. Integration may require only parts of existing theories. In such a case, the needed components are combined into a new theory, which founds a new understanding, in contrast to combinations of entire theories in additive integration.

key physiological features of organisms that might control nutrient cycles and combines them with geochemical perspectives of the balance among elements in the environment. Finally, the prospective joining of patch dynamics, metacommunity models, and island biogeography theory mentioned in Chapter 6 (Box 6.2) is likely to occur via an extractive process, with some of the components of the constituent models being retained while others are dropped. For example, rates of extinction and colonization may become phenomenological characteristics produced by dispersal and metapopulation dynamics.

Both additive and extractive integration are likely to involve (1) progressive sharing of empirical and theoretical contents between domains or (2) finding a linking relationship. For two or more areas of research to be successfully integrated, the number of definitions, concepts, or models (Chapter 3) they have in common must be increased. Integration will further be enhanced if the shared conceptual devices are fundamental to the areas to be integrated. For instance, basic predator/prey and competition models share many mathematical conventions and assumptions about the closed nature and structure of the populations and habitat, the dynamic nature of population growth, the role and representation of time lags, and so on. Finding a linking relationship may permit the two areas to retain many of the differences due to their specific foci, but still to simultaneously examine effects of competition and predation on community structure, stability, diversity, and so on. In limnology, productivity of freshwaters has been well integrated with plant physiology by the successful application of the "law of the minimum." This law codifies the observation that when one factor is limiting, then others are unlikely to be so at the same time. Here, the two research areas were linked by the discovery of the limiting role of phosphorus (e.g., Rigler 1982).

These two kinds of integration can occur at any scale or breadth of scope, but scope can be important in determining the significance of integration. Strictly speaking, integration may even occur at the fine scale by combining models that are relatively near one another in focus and approach. Such fine-scale integrations are part of the daily work of ecologists. In contrast, and perhaps in closer conformity with the usual connotation of the word, integration may address broad arrays of diverse ecological phenomena. Ultimately, one may hope for a grand integration or unification of all the disparate approaches to ecology. Some commentators propose that ecology has been treated as a single discipline, not because of true theoretical integration but because of habit and accidents of interest among key founders of the discipline (Hagen 1989). If this analysis is true, it represents an underlying powerful motivation for integration among domains and paradigms in ecology. We suspect that a discipline that is integrated by more than historical accident and habit would be better able to articulate and conduct integration.

The definition of extractive integration raises the qustion of how to avoid arbitrariness in its application. Of course, exclusion of certain aspects of the precursor theories would only mean that those components were not needed in the new theory (Darden 1991). However, it would not mean that components that were potentially contradictory within the desired new understanding had been excluded arbitrarily. Exclusion of components of areas contributing to a new integration should therefore indicate only that those components are not relevant. If a component of an existing theory is shown to be incorrect during the attempt to synthesize an integrated understanding, that component would have to be replaced or corrected in the contributing theory as well as in the integration. One of the critical safeguards of objectivity in science is that the components of unified understanding and, indeed, independent successful theories on related phenomena not contradict other accepted and well-confirmed theories relevant to the domain (Lloyd 1988). Contradictions between theories of subjects that ought to be related suggest the opportunity to develop deeper theories that might expose the unity among the phenomena. This is the case in the theories of the four fundamental forces in contemporary physics (Joseph 1980, Rohrlich 1987).

Integrations in ecology, especially those of broad scope, may often be hierarchical. The hierarchical structure of several broad theories relevant to ecology — for example, evolution and succession (Pickett et al. 1987) — suggests that such a structure would also characterize novel integrations. This view is reinforced because the hierarchical view of ecological entities and processes is such a pervasive and successful one (Allen and Hoekstra 1992, O'Neill et al. 1986, O'Neill and King 1998). Similarly, competition, predation, species tolerances, and habitat structure have been combined into one hierarchical model that predicts general species abundance and specialization trends (Kolasa 1989, Kolasa and Romanuk 2005, Kolasa and Waltho 1998) in abstraction from the specific factors determining the individual species performance. The model adopts a simple common denominator, that of habitat suitability, which permits bypassing those specific factors that have been treated as separate areas of inquiry and that produce a high degree of contingency. These include competition, predation, physical environment, disturbance regime, and many others. In our terminology, the model supplants, for the sake of community structure analysis, a range of mechanisms with a single more general mechanism.

New explorations of ecological boundaries as integrating phenomena across various scales and ecological disciplines have also been described hierarchically (Fig. 7.3). The integration identifies a fundamental process, flux of material, organisms, energy, or information across heterogeneous space (Cadenasso et al. 2003b). This phenomenon is hierarchically broken down into contributing phenomena and processes: (1) the nature of the flux, (2) the contrast between the elements that make up the spatial heterogeneity, and (3) the architecture and arrangement of boundaries or gradients between elements (Cadenasso et al. 2003b). Each of the contributing processes would require specification of scale, source-sink relationships, sensitivities to control, and so on. These features would constitute the details of specific quantitative or experimental models. For example, a heterogeneous mosaic can be identified to span forest-field boundaries. The flux of seed rain across the boundary might be chosen as the transfer of interest, and the assumption made that the three-dimensional architecture of the forest-field edge, with its dense display of branches and foliage of canopy trees, understory trees, and shade intolerant shrubs, would control the seed flux from forest to field. In experiments designed to test this assumption, Cadenasso and Pickett

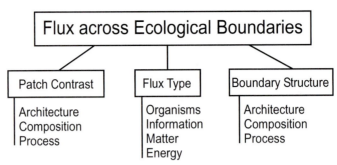

Figure 7.3   A causal hierarchy for the structure and function of ecological boundaries. This hierarchy can apply at a variety of scales and kinds of systems. For example, it can apply to terrestrial-aquatic boundaries, soil-atmosphere boundaries, or the boundaries between contrasting vegetation types. It indicates that models and hypotheses of boundary structure and function can consider three main realms: (1) the contrast between bounded, adjacent patches, (2) the kinds of fluxes across the boundary, and (3) the nature of the boundary itself. Modified from concepts in Cadenasso et al. (2003).

(2001) discovered that removing the drape of leaves and branches at the edge of the forest did in fact significantly increase the flux of the seeds of light demanding field species into the forest interior distant from the edge. One can imagine still more detailed models than this experiment, which would expose finer scaled mechanisms of wind flow and resistance to the load of carried seeds. Other work showed the effect of the forest-field boundary (1) on the environmental gradients between the forest and the field (Cadenasso et al. 1997), (2) on animal movement and impact on tree seedlings (Cadenasso and Pickett 2000), and (3) on the atmospheric deposition of nutrients (Weathers et al. 2001). Each of these more detailed studies fleshes out the lower level branches of the theoretical hierarchy aimed at explaining fluxes across heterogeneous space. The theoretical hierarchy, having been developed to promote integration, can be applied to transfers across boundaries as different from those within the soil, and between soil and atmosphere, between land and streams, or between biomes, for example (Cadenasso et al. 2003a, 2003b).

Reductionism is a persistent problem that is relevant in the context of ecological integration and the hierarchical structure of theory. Reductionism is unfortunately a loosely used and often disparaging term in ecology. However, philosophers of science take it to mean several things without prejudice. First, reductionism means that the material bases of higher level phenomena are those phenomena and structures of lower levels of organization in nature. This is simply the acknowledgment that there are no mysterious vital forces that need to be invoked in moving from physics and chemistry to biology. This viewpoint is entirely appropriate and is considered one of the basic assumptions of contemporary science (Price 1961). A second philosophical meaning of reductionism is that the theories of a science applicable to a higher level of organization in nature should be strictly and formally reducible to the theories of lower levels. There has been much debate about the possibility or desirability of reducing biology to chemistry and physics (Mayr 1982). This kind of reductionism is a holdover from the days of the now-displaced philosophy of logical positivism, which considered theories to be deductively linked series of statements (Chapter 1). By and large, this problem has been abandoned as unproductive and misguided (Rosenberg 1985), although the arguments echo even now. Modern philosophy recognizes that the families of models appropriate to the various large topic areas in science can have productive, independent lives. Of course, the conceptual components and generalizations of ecology cannot contradict accepted theories of lower level sciences, but they need not be formally reducible to them. For example, thermodynamics and the conservation of matter are necessarily basic assumptions of ecosystem theory (Reiners 1986), but they do not constitute ecosystem theory while a number of postulated biological rules might do so (cf. Jørgensen 2002).

Novel attempts at integration in ecology have often looked upward in scale or level. The landscape perspective illustrates integration upward at a coarse scale. Whereas community and ecosystem ecology have typically focused on relatively homogeneous stands or patches (O'Neill 1999), landscape ecology recognizes that individual systems or stands have a spatial context of similar and contrasting patches in which they are embedded and function. The appearance of the relatively new, self-identified discipline of landscape ecology (Forman and Godron 1986) has stimulated the search for appropriate measures of coarse-scale heterogeneity and pattern, the detection of functional connections between components of landscapes, and the constraint by landscapes on lower level entities such as populations and individuals (Holland et al. 1991). In addition, how the interactions of lower level entities may determine the structure and function of landscapes as a whole is a contemporary research subject (Lovett et al. 2005, Risser 1987, Turner 1989, Weins 2001). These concerns show how an upward integration can contribute to the growth of understanding.

It is important to recognize that although the integration represented by landscape ecology started as a scaling up, now that it has been established, there are other ways in which it has been generalized. Once the fundamental kernel of the reciprocal effects of spatial pattern and ecological process was articulated (Turner 1989), the door was open to apply this nugget to any scale. Spatial pattern is a phenomenon that appears on any scale ecology wishes to address. New developments in the area of metacommunity ecology show conceptual affinities as well are relative strengths to vis-à-vis landscape ecology (cf. Holyoak et al. 2005). Combining the two areas may be possible and profitable. Given that metacommunity ecology is potentially unifiable with patch dynamics and island biogeography, the prospects for larger unifying theory are not as pessimistic as ecologists may sometimes think (cf. Taylor and Haila 2001). Therefore, the fundamental question of what is the effect of spatial pattern on ecological process belongs to any scale, not just the so-called landscape scale. To recognize this scale-independent core of landscape ecology, Allen and Hoekstra (1992) suggested that landscape was better conceived as a "criterion of observation," rather than a specific spatial and temporal scale. A criterion of observation is a lens or filter through which studies are framed. Addressing flux across heterogeneous space or the effect of pattern on process and process on pattern can be done at any scale. Because of this, the specification of the scale(s) chosen for a particular study or model is critically important. Landscape as a fundamental concept does not itself specify a scale. Taking landscape as a criterion of observation extends its integrative power downward, as well as in the upward direction it originally possessed.

The recognition of a new level of organization was also important in the progress of classical genetics from its relatively primitive state in the early 1900s to its mature state 20 years later. The new level of linkage group was a critical upward integration of genes that was required to understand the functions of the original lower level (Darden 1991).

The hierarchical perspective makes the persistent debate between holism and reductionism moot. The modern view of nested hierarchies of phenomena, entities, and related theories suggests that any legitimate scientific question can be aimed at any focal level, as long as that level is clearly specified and the causal links to other levels are recognized. To understand context, constraint, and mechanism, investigators then examine adjacent levels. Questions couched in terms of "why" require an upward, and often longer or larger scale, examination of context. Questions couched in terms of "how" — that is, by what mechanism — look downward at least one level. However, there is no a priori need to look to the lowest possible level for mechanism. Furthermore, knowing when to stop provides an important safeguard against knowing too much about too little. Indeed, knowing where to stop for an adequate causal explanation is one of the principal requirements for success in science. Without well-considered downward limits, a practically endless reductive spiral might occur. At the least, an inappropriately deep reduction to entities considered to be the real units of one of the lower level sciences might be tempting (Longino 1990). For example, attempting to explain all ecological phenomena at the molecular level because those units are considered real would leave many ecological phenomena unexplained. Even the explanation of such clearly molecular-based processes as whole-plant photosynthetic assimilation cannot proceed without adding knowledge of canopy architecture, source-sink relationships within plants and between plants, soil and atmosphere, effects of plant consumers, and resource availability, to name but a few. Events and processes at distant lower levels are rarely of direct impact on a focal level. In fact, according to hierarchy theory, their expression declines with their distance downward from the focal level (Allen and Hoekstra 1992).

III. Questions for Integration

In the previous chapter, we identified the fundamental questions that advance understanding within a subject area to be those that deal with the completeness, soundness, and establishment of theory. Is it possible to identify the kinds of questions that might drive integration *between* different subject areas? We believe the kinds of fundamental questions within an area — which drive establishment, refinement, rejection, replacement, and extension in scope of a theory or theoretical component — are likewise applicable to integration as well. Using the parallels with integration within an area, fundamental questions aiming to increase integration *between* subjects might (1) suggest the need for a new integration, (2) promote the development of theories or missing components of theories to support integration, or (3) lead to the exclusion from the integration of irrelevant components of the source theories.

As an example of a fundamental question promoting integration, we present a problem in competition theory. The phenomenon of apparent competition (Holt 1987) invokes mechanisms outside classical competition theory. Apparent competition requires that indirect effects that act through animals that consume the competitors be accounted for, as well as direct effects of resource exploitation by the competitors. Integration thus calls for a theory that deals with all these mechanisms. It calls for a theory that deals with both competitors themselves and their consumers. Another example of the need for fundamental questions comes from the ecosystem energetics area that we discussed earlier. Müller (1997) presented ecosystem theory as an aggregate of various theories in need of integration. A question about links between ecosystem development and energy flows illustrates the many possibilities arising within the area. Self-organization and buildup of species diversity that accompany ecosystem development have predictable consequences for the partitioning of energy flows in terms of structure, efficiencies, and dissipation. An integrated theory would thus need to accommodate the processes of adding and replacing species within an ecosystem in the context of changing energy flow structures and efficiencies.

Fundamental questions concerning integration would explicitly focus on the coarse scale or on phenomena of broad scope and applicability, such as the ones dealing with competition or ecosystem energetics just described. A fundamental question that specifically incorporates a broad spatial or temporal scope within its purview, regardless of whether it is truly integrative or not, might be called informally a "big question." For example, asking how terrestrial ecosystem processes are linked with atmospheric processes at the subcontinental scale applies to the entire Earth and, hence, is a big question. Incidentally, it is also a societally important question.

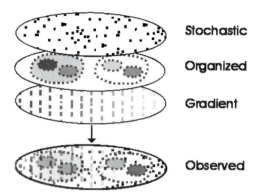

Figure 7.4   Observed ecological patterns and their common components. At any scale, observed ecological patterns are a composite of stochastic influences (e.g., weather, history, transient organisms, random colonizers, or propagules), gradients (intensity of systematic factors, frequency of stochastic factors, mean values of any biotic and abiotic variables), and deterministically organized processes (predator prey, mutualistic, or competitive interactions, phenology, migrations, ecosystem processes). One desirable feature of a unifying theory will be the ability to interrelate these patterns. For example, as shown in the figure, organized processes are more pronounced at one end of the gradient while suppressed at another. An actual theory might be able to give specific expression to such relationships.

Since so many pressing environmental problems deal with regional to global scales, big questions are often likely to be societally important. Integration of ecology and other disciplines at such coarse scales is also likely to be quite important. However, it is important not to confuse the spatial or temporal scope of a question with its power to integrate across subject areas.

Another dimension along which integration could occur pertains to the need to resolve the origin of ecological patterns. Theoretical approaches that help clarify the nature of complex patterns (Fig. 7.4) in a unified manner will make considerable progress in bringing together now largely separate areas of ecology such as macroecology, ecosystem ecology, community interactions, fire ecology, patch dynamics, and others.

IV. Radical Integration and Paradigms

The previous section considered how integration across domains relates to the structure of and change in understanding. That discussion was largely focused on domains of understanding within ecology encapsulated in theories that are the day-to-day concerns of ecologists. However, there are other aspects of integration which are not obvious in such a focus. In this section, we focus on the nature and problems of integration across the most extreme contrasts that can exist within ecology. Because ecology can range from geochemistry at one extreme to physiology and genetics at the other (Fig. 1.1), the contrasts across which integration might occur are indeed vast. Along the spectrum of ecological specialties, the contrast is so large that it suggests contrasting paradigms.

The term "paradigm" is an extremely common and important one in modern science and philosophy. It has captured the imagination of scientists who are eager to lead a revolution and overturn a dominant paradigm, thus gaining fame (and maybe even a little fortune through grants). Hence, the term "paradigm" has become a hopeful part of the daily parlance of science. Kuhn's (1962) early use, however, was quite multifaceted, and unclear to many readers; in a later edition (Kuhn 1970), he clarified the term considerably: *A paradigm is the worldview, belief system, series of assumptions and techniques, and exemplars for problem solution held in common by a scientific community*. The worldviews held by population ecologists, physiological ecologists, ecosystem ecologists, landscape ecologists, and evolutionary ecologists, to name a few, are disparate enough to suggest paradigmatic differences within the science of ecology.

Even the refined definition of paradigm provided by Kuhn can be interpreted — in the extreme — to suggest that changes in scientific understanding are subjective and culturally relative. A related view of the sociological basis for change in science (Feyerabend 1975) can be considered rather anarchistic. However, in dealing with integration, there are two reasons to eschew a purely subjective, relativistic view of paradigm. First, in the broadest sense, a paradigm is the largest theory held by a scientific community (Kuhn 1977). However, the narrow connotation of the term "theory," which was dominant in philosophy when Kuhn first wrote, would not have suggested the breadth and pervasiveness of viewpoint that he required to analyze the great revolutions in science. Indeed, that narrow and inappropriate view of theory is still held by many scientists. We suggested earlier how that narrow view — the statement view — is inappropriate. The contemporary view of theory as a system of conceptual constructs within a specified domain (see Chapter 3) would, we suspect, have proven useful to Kuhn in describing major conceptual turnovers in science. Therefore, paradigms have an objective side on which we will focus in the remainder of this chapter. Second, Kuhn (1970) made important assumptions about the nature of the scientific community, which led to the conclusion that subjectivity plays a big role in theory choice. Subsequent analysis of the structure and functioning of scientific communities indicates how they can, in fact, operate objectively (Longino 1990; Chapter 8). Hence, in discussing integration across paradigms within ecology, no assumption of subjectivity or cultural relativism should be made.

Before exploring the paradigms that exist within ecology, we must expose a hidden complexity in the concept of paradigm. A hierarchy of paradigms exists within ecology. Of course, at the largest scale, the assumptions and metaphors of the society at large affect ecology. The largest societal worldview is essentially invisible to the practice of science, but we will mention how certain of its aspects have become visible and, as a result, how their impact on ecology has changed over the past several decades (Chapter 8). The second scale of paradigms that affects ecology is that of the background assumptions of science as a whole. The assumptions of contemporary science appear clearly only in contrast to those of medieval or ancient science. The next level of paradigm is the one that applies to ecology as a whole. Certain key aspects of the all-ecology paradigm are indicated by the definition we cited earlier for the entire field (Chapter 1). But there are still more kinds of paradigm relevant to ecology than the all-inclusive ones. Nested within the broad science of ecology are specific ecological disciplines, each of which has its own paradigm. We will use this hierarchical scheme to differentiate paradigms of contrasting breadth and inclusiveness, and we can use specific terms to highlight the differences (Box 7.1).

A strategy that we use to facilite the discussion on further integration is borrowed from community ecology. This strategy resembles ordination, a suite of techniques aimed at extracting and reducing the number of dimensions that affect a complex, multidimensional phenomenon.

A. Disciplinary Paradigms in Ecology

Here we highlight the different viewpoints, approaches, and exemplars that characterize the various broad specialties within ecology. The paradigms of ecology lie along two axes. The first axis identifies a focus on discrete object-like entities versus a focus on continuously variable quantities. We call this axis Things-Stuff. The second axis of contrasting paradigms separates a focus on entities whose history is a critical ingredient of their behavior from a focus on entities whose behavior is determined only by their current state. This second axis we call Then-Now.

BOX 7.1 Nesting of Paradigms Relevant to Ecology

The paradigms that apply in or encompass ecology are part of a nested set that extends from the whole of contemporary science to the specialties within ecology. We identify three nodes as important here:

Scientific paradigms. Relevant to all modern science: materialism, causality
Ecological paradigms. Relevant to all of ecology: nonequilibrium or equilibrium, multiple or simple causality, unilateral or reciprocal control
Disciplinary paradigms. Define disciplines within ecology: population or ecosystem, instantaneous or evolutionary

---

The terms we use for the axes are informal, perhaps amusing, and, we hope, memorable. However, the contrasts are serious and sometimes a problem for ecological unity.

We will first examine the contrast between the ecological paradigms for Things and Stuff. A more formal terminology would be entities versus quantities. The epitome of a discrete Thing or entity is an organism. Ideal organisms come in whole units enumerable by integers. They have clear boundaries, and the genetic, developmental, structural, or behavioral differences among individuals are recognized as key to understanding assemblages of organisms. Stuff or quantity, on the other hand, exists as a pool or aliquot that can be described by continuous variables in the form of real numbers. If the pool comprises individual items, their size is so minuscule relative to the size of the entire pool that any idiosyncrasies of individual items are not important in describing the behavior of the pool. In other words, individual qualities disappear, and the behavior of the entire ensemble claims the attention of researchers. An ideal example of Stuff is any chemical or ion found in the natural environment. Carbon dioxide, as Stuff, is described by volume, mass pressure, flux, and so on. The characteristics of individual molecules are unimportant for describing and working with the flux and behavior of the gas, in much the same way that the gas laws do not require an understanding of the location or speed of individual molecules in order to apply. Thus, the ordination axis can describe gradients between Things and Stuff as well as distinctions between them.

To follow a recent trend, we hurry to assure the reader that this distinction depends on scale of description. "Things" such as plants may become "Stuff " in global carbon models and algae can also be treated as "Stuff" in a lake model. Likewise, particulate organic matter (POM) may become objects with diverse properties for someone studying feeding behavior of blackflies in a stream. Like so many choices ecologists have to make, whether a phenomenon is resolved as Things or Stuff depends on scale. Stating the scale of discourse thus becomes important in recognizing these paradigmatic differences.

1. Thing Ecology

Studies of ecological entities whose individual enumeration, characteristics, and behavior are important include such areas as plant and animal demography. In such studies, the origin and demise of individuals are important; their rates of growth and mortality and their genetic, size, and behavioral differences are used as explanatory or predictive variables. The genetic and behavioral differences among organisms make it possible to examine their roles in evolution.

Indeed, evolution requires the existence of individual entities that have the same characteristics as ideal organisms. Integrated groups of organisms such as populations or communities can also be treated as entities. Metapopulation and metacommunity ecology rely on such conceptualizations. Likewise, disturbed patches in a landscape can be considered individuals. They have birth and closure rates, sizes, spatial distributions, and internal heterogeneities. These characteristics are important for understanding how organisms respond to them, how landscapes are structured, the role of disturbance in natural and anthropogenic environments, and so on (Levin and Paine 1975). Perhaps even species or evolutionary lineages can, for certain purposes, be considered individuals (Eldridge 1985, Hull 1970), but that debate is one we will not settle here.

2. Stuff Ecology

Ecological studies that focus on pools and fluxes of materials and energy fall into the category of Stuff ecology. Ecosystem energetics, metabolic studies within large habitats, and nutrient flow studies are all cases of research on ecological Stuff. Typically, a budgetary approach is required to understand fluxes, so system boundaries, input rates, flux rates, and the controls of fluxes and transformations must be combined into a mass balance (Box 7.2; Likens et al. 1977, Schlesinger 1991). For example, the study of phosphorus dynamics is motivated by its role as a limiting material in aquatic ecosystems. Such studies require that factors such as the following be determined: the form the nutrient takes, how much of each form is in each compartment as well as the whole of the ecosystem, how fast it is transformed within various compartments, how much moves between compartments per unit area per unit time (Fig. 7.5), and how much enters and leaves the whole ecosystem per unit time. In addition to predicting the changes in amount and species of nitrogen over time and space, the nature of controls on the fluxes must be discovered. It may be desirable to know the relationships between fluxes of different nutrient elements and the relationship to the flow of energy in the system. Ultimately, the discovery of ecological laws that govern the fluxes of matter and energy in ecosystems is a goal. This task requires that the caveats invoking the issues of entity and ecosystem definition in Chapter 3 are successfully tackled.

The contrasts between the Things and Stuff paradigms, although briefly drawn, should suggest that the resulting understanding in each of these paradigms will take different forms. How the systems of interest are described, delimited, and modeled differ. For example, Stuff ecology often employs system components as black boxes in which the detailed interactions among organisms are hidden, for analytical convenience, whereas Thing ecology often explicitly addresses the differences among individuals, as in size- or age-structured demographic models (Silvertown 1982), and takes many of the fluxes of Stuff as a given. The numbers of replicates for studies within

---

BOX 7.2 Approach of Stuff Ecology

- Determine or set boundary of system (e.g., watershed, field, landscape element, microcosm, arbitrary).
- Assemble model of components and pathways of flow in the system and in and out of the system.
- Measure the pools of materials or energy in each component.
- Measure the rates of flux between components and in and out of the system.
- Determine the processes controlling fluxes.

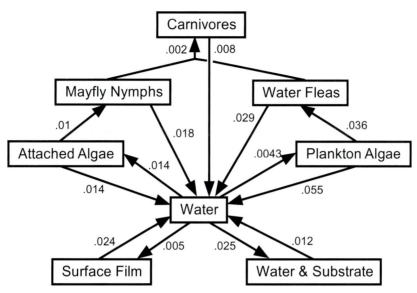

Figure 7.5   Phosphorus flux in an ecosystem provides an example of the use of the Stuff paradigm in tracing the amounts of phosphorus in various components of an ecosystem, and the rates of flux between components. Modified from Whittaker (1961). Ecological Monographs. © Copyright 1961 Ecological Society of America. Reprinted by permission.

each paradigm may differ for practical reasons, since there are fewer boundable ecosystems on Earth than individual organisms of most species. The kinds of variables measured and the kinds of controlling processes sought will differ greatly as well. For instance, in Stuff ecology, variables such as temperature and moisture are important in understanding processes, whereas in Thing ecology, the evolutionary history or optimal allocation of assimilated energy is especially significant. Finally, what counts as a significant question under each paradigm will also differ. A typical question in the Stuff paradigm might be, "How are the productivities of adjacent but contrastingly structured ecosystems linked?" In the Thing paradigm, a typical question might be, "What regulates the numbers of species in fluctuating environments compared with constant environments?" Clearly, the research tactics and the structures of theories must differ under these two paradigms. Even within the same discipline, paradigmatic divisions and contradictions can persist and reflect differences in traditions and subject matter, just as they do between plant and animal ecologies (Austin 1999). Therefore, the taxonomy of theories in these different paradigms would also differ (see Chapter 5). However, difference does not imply qualitative disparity. The two paradigms do not stand in a superior-inferior relationship. They only address different subjects.

The second axis contrasts the paradigms of Then ecology and Now ecology. The viewpoints of this axis differ depending on whether or not history is considered to have a role in explaining the phenomenon of interest.

3. Then Ecology

The basic assumption in historical disciplines is that some past state(s) of the system of interest affects the current state. In ecology, this viewpoint is expressed in a variety of ways. One, evolutionary ecology, takes heritable differences and the principle of allocation, or trade-offs

BOX 7.3 Principle of Allocation

Assuming that organisms have a limited assimilated energy or material pool, matter or energy used for one structure or function cannot be used for a competing structure or function. Assimilated matter or energy must therefore be traded off between competing functions.

(Box 7.3), as the integrated realizations of core historical mechanisms and constraints (Cody 1966). Responses that are constrained by limited assimilated energy or materials, and that are heritable, may experience natural selection and therefore may influence the current or future responses of the organisms or the assemblages of which they are a part. Expected responses of organisms can be generated by the assumptions that genetic variation and selection are unconstrained, yielding an idealized behavior against which actual behavior can be gauged to assess the impact of other biotic or abiotic factors. The articulation of evolutionarily stable strategies or the proposition of coevolved communities with reduced competition exemplifies this research tactic.

A second manifestation of the historical paradigm is a mechanistic one. It can appear in paleoecology, in which the past states of vegetation, species, or ecosystems can be sought by a combination of retrospective studies, modeling, and simulation based on physiological or autecological information on the species (Davis 1983, Shugart and Seagle 1985). Historical ecology can be used to forecast future responses of ecological systems and to explain or to forecast spatial differences in system structure or function. Historical ecology is becoming increasingly important, especially as the role of people and civilizations in the past is becoming better understood (Cronon 1983, Foster et al. 2003, Williams 1989) and as ecologists try to forecast the impacts of human-induced climate change. Then ecology can thus include the future as well as the past.

General issues that typify historical approaches to ecology include the persistence of system conditions; the role of episodic events; the significance of physical disturbances to communities, populations, and ecosystems; and the nature and impact of ecological memory (Facelli and Pickett 1990). Time lags, contrasting initial conditions, echoes of the past, and priority effects are examples of conditions that require a historical approach (Fig. 7.6; Pickett 1991). The storage effect (Chesson 1986) is an example of an emerging approach to historical ecology. The storage effect recognizes the generality of persistent life history stages or size classes established during transient events as factors in community and system organization. An excellent discussion of the historical paradigm, written by a historian but remarkably applicable to ecology, can be found in the book by Gaddis (2002).

Succession theory is a community-based expression of the Then paradigm. Here, a sequence of arrivals and a sequence of habitat modifications become of primary importance in structuring and interpreting ecological phenomena. Without the time perspective, understanding would be impossible to achieve. Not only is the basic idea of succession a case of Then ecology, but so are its historical refinements. Succession studies have exposed legacies of past land use, ideosyncratic events of invasion or disturbance even within different successional trajectories.

Finally, interesting hybrids between the evolutionary and the historical Then perspectives can be found. Examples include the concepts of the extended phenotype or organism (Dawkins and Dennett 1999, Turner 2000), niche construction (Odling-Smee et al. 2003), and ecosystem engi-

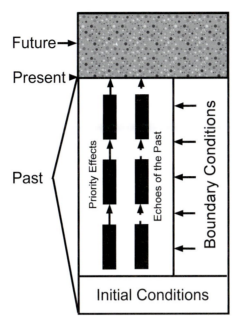

Figure 7.6 Aspects of the past. The past refers to those events and interactions that occurred prior to the present, regardless of whether they left a record or history. The past can influence the present (indicated by the threshold line between past and future) as echoes, priority effects, and initial and boundary conditions. Echoes are past events whose effects have not been constantly felt on a system but appear sporadically. Priority effects are those that result from different orders of events in time. Boundary conditions are those that constantly affect and constrain the process of interest, whereas initial conditions are those that set the starting state of the system. All such effects may persist through time and may determine, at least in part, the contemporary status of the system. Adapted from Pickett (1991). Reprinted by permission of John Wiley & Sons, Ltd.

neering (Jones et al. 1994, 1997). Organisms can create their own abiotic environments, which often persist, affecting the organism that created the environment and its offspring (e.g., an animal burrow). The capacity to construct such abiotic environments resides in organismal traits that are, of course, subject to natural selection. Although the abiotic environment obviously cannot evolve, it can nevertheless express semantic information inherant in these organismal traits (e.g., the burrow). At the same time, the persistence of these structures and their feedback effects on the organism and its offspring can be considered as an independent historical legacy.

4. Now Ecology

If certain ecological processes can be assumed to be instantaneous in their workings, then the focus of research can be strictly contemporary. Billiard ball physics is the epitome of an ahistorical discipline, in which information on the previous conditions of the systems adds nothing to the information on the current state of the system; instantaneous measures of the velocities and positions are sufficient predictors of the future state. In fact, velocity as a variable integrates and obviates the variables of speed and direction of billiard balls that are the historical components of the system. Within ecology, environments, populations, communities, and ecosystems that have a low capacity to retain states or their effects through time or that adjust quickly to environmental changes, reasonably can be the focus of the instantaneous Now paradigm.

Examples of instantaneous ecological studies are those that examine niche partitioning of quickly responding organisms in contemporary time. Behavioral ecology can often safely focus on strong instantaneous interactions whenever past learning experiences can be discounted.

As usual, much of the distinction will depend on scale. For example, a predator/prey interaction is largely determined by the current state or Now perspective. So understanding what happens, or what happens next, requires the knowledge of population densities, encounter frequency, attack success rates, the energetic requirements of the predator, and the reproductive capacity of both. However, understanding of a predator/prey cycle may require much longer term knowledge of time delays, spatial refugia, and stabilizing processes. Some model cycles stabilize after a number of generations, and each time step (generation) must be known in order to project the next. In such situations, a predator/prey system drifts, at least in some aspects, toward the Then suite of conceptual tools. Motivated by similar considerations Berryman (1987) appealed for the rejection of the dichotomy and for the focus on developing an integrated theory based on relationships, principles, and processes to supplant the equilibrium and nonequilibrium poles. More recently, Wu and Loucks (1995) exposed virtues of the nonequilibrium approach. While such a unification may be feasible, not all areas of ecology may be amenable to a theoretical fusion of Now and Then perspectives.

The contrast between the historical and the instantaneous paradigms appears not only in ecology but has parallels throughout science and other fields. In economics, decisions are made at the margin, essentially assuring that an instantaneous assessment of costs and benefits is sufficient (Hall 1993). Aspects of the relatively new field of environmental economics make different assumptions about the roles of past resource use and future costs compared to classical economics. In science, Gould (1986, 2002) was a major proponent for the historical approach. Historical sciences have received short shrift in the history and philosophy of science. The favorite models for classical philosophy of science — mechanics and logic — are tacitly ahistorical disciplines. The methods of inference in the historical sciences, involving as they do multiple lines of inference and retrospective studies (Likens 1989), have hardly been acknowledged, let alone well explained by the simplistic "kick-it, predict-it, measure-it" model of hypothetico-deductive experimental science (Chapter 2). As with the contrast between the Things-Stuff paradigms, instantaneous and historical ccologies are of equivalent quality. They simply focus on systems having different time scales of essential processes, make different assumptions, and therefore require a different mix of methods.

The recognition of these differences is not to be taken as their permanent sanctioning or acceptance of the intrinsic differences among processes or entities. Much may depend on the quantitative separation between ecological and evolutionary or historical time (cf. Cooper and Ruse 2003). Indeed, our message throughout the book is that theoretical integration serves understanding. Consequently, we deem it possible and hope that some of the polarities that characterize today's ecology will be replaced by more unified components of the disciplines' theoretical framework. We can envision a transition between at least some of the historical and instantaneous models of ecological phenomena by considering a couple of examples (Box 7.4).

B. Misapplication of Paradigms

The four paradigms that appear within ecology have not been explicitly recognized throughout much of the science. Their legitimate and necessary differences in focus and mix of methods have received little attention, yet these paradigms are such major ways to look at the world that they are pervasive. However, without analysis, it is possible to misapply them.

For example, the Things-Stuff paradigms have been confounded in their application to ecosystems. The idea that ecosystems evolve in a way that is parallel to how populations evolve is

BOX 7.4 Examples of Instantaneous Processes with Historical Modifiers

Genetic

Zooplankton diurnal vertical migration has been found to undergo modifications using a genetic mechanism in response to fish predation (Gliwicz 1986). Specifically, in lakes stocked with fish, intense predation causes zooplankton to migrate to deeper and darker waters during the day and to move to shallower feeding zones at night. The magnitude of the effect depends on the number of years zooplankton populations were exposed to predation. In this situation, any dynamic, instantaneous model of zooplankton grazing on phytoplankton would fail due to evolution of behavior under predation. However, another model that incorporates the history of the population exposure to predation might succeed.

Learning

While anecdotal, the example is interesting and makes the point clearly. In some areas of Africa (for example in Tsavo, Kenya), lions show higher incidence of attack on humans than in others. A credible explanation for this behavior is "tradition" (Peterhans et al. 2001). Once lions learn that a human can be successfully treated as prey, this knowledge is passed from generation to generation. There is nothing unusual about this, as lion prides are known to specialize in various types of prey including zebras, buffalo, and even elephants, and such specialization is pride-specific over many generations. Thus, a Now perspective of predator-prey models would be strengthened if learning and the ensuing change in attack and success rates were incorporated into the models. The task may be technically difficult but is legitimate and would result in an increased integration of behavior and predator-prey theory as well as a more refined understanding of the observed diversity of behaviors.

   Both examples show that a historical explanation can, in principle, converge into a Now mode of interpretation if the subtle historical effects are recognized (via historical and comparative studies) and then appropriately incorporated into the Now perspective.

an error. Certainly ecosystems change or develop, and their history is important to their current and future states and dynamics, but ecosystems cannot evolve in the Darwinian sense, because they contain abiotic entities, because they are not unambiguously bounded and unitary, and because they do not have heritable characteristics among which selection can discriminate. Some developmental models may help describe ecosystem changes (e.g., Ulanowicz 1986), but this cannot be Darwinian evolution. In addition, separate research traditions have different views on ecosystem development as exemplified by differences between the Russian and Western ecosystem paradigms (Lekevicius 2003). Such differences may result in confusion but also may carry some benefits by presenting and sharpening conceptual challenges. In contrast, species associations, the focus of community ecology, are sometimes viewed as evolving entities (e.g., Wilson 1989) and such an interpretation appears to be consistent with evolutionary theory in the broad sense.

   Another and a rather frequent misuse of "paradigm" is the use of the term to label a set of new findings, change in focus, or new models. Such misuse may reflect genuine belief by a group of ecologists that the new developments are as deep as to change "the worldview, belief system, series of assumptions and techniques, and exemplars for problem solution held in common by a scientific community." However this belief is not automatically equivalent to a wide consensus as to their meaning. For example, an increased interest in the effects of species diversity on ecosystem function has some elements of change that ultimately might suggest a paradigm shift or an integration across paradigms (cf. Naeem 2002) but does not in itself constitute an altered paradigm. Specifically, the view that diversity governs ecosystem functioning is not widely held but represents a working hypothesis, particularly in the context of already known factors that govern productivity, trophic structure, nutrient cycling, or respiration. The available evidence is being hotly debated, and it certainly is not a common and shared worldview among ecologists (cf. Duffy 2002). A hasty acceptance of a mislabeled process as a change in paradigm has the potential for interfering with a balanced pursuit of legitimate and promising questions.

Another problem with paradigms occurs when internal contradictions and inconsistencies between subgroups of practitioners remain undetected. An example comes from community ecology where plant and animal ecologists separately developed a host of theoretical and empirical tools for describing and analyzing the structure of communities (Austin 1999). Plant ecologists focused primarily on individual species tolerances and species replacements along spatial and temporal gradients as a function of those tolerances. In contrast, animal ecologists pursued the development of community ecology around species interactions such as competition and predation. While the two fields borrow from each other, the communication between these two areas is slow (Austin 1999). Consequently, models and theory capable of the comprehensive treatment of both plant and animal diversity may have been delayed. Ultimately, most scientists hope to achieve a degree of theoretical unification and treat, correctly, the pluralism of paradigms as necessary but temporary (Mitchell 2002).

C. A Focal Point for Integration

An ideal ecological worldview emerges from identifying the two axes of paradigm contrast in the science — Things and Stuff and Then and Now, or entities and quantities and historical and instantaneous (Fig. 7.7). The two axes can be conceived of as overlapping and as specifying at their intersection an integrated view of how the world works. We consider the bull's eye at the center of the universe defined by the four paradigms as a potentially complete view of any ecological entity or process. Like the proverbial elephant encountered for the first time by the blind, ecology seems to be a science of fluxes, or a science of organisms, or a science of instantaneous adjustment, or a science of historically important environments and interactions. In reality, it is a science of all these things, and the four paradigms are simply different perspectives of the ecological elephant that exists as Things and Stuff in the Then and Now.

The recognition of the four disciplinary paradigms in ecology suggests an opportunity for a grand integration in the field. These four paradigms are usually pursued without much reference to one another. Unfortunately, what cross reference exists is usually disparaging. However, each is a legitimate perspective of the ecological elephant. The chance to construct an integrated model of the entire elephant is, therefore, an exciting possibility not to be neglected.

To exploit the target of cross-paradigm integration, we need four highly developed areas. Unfortunately, some are in a surprising state of conceptual infancy, for example, ecological individuals or entities. We have proposed a general framework for identifying and working with ecological entities (Kolasa and Pickett 1989) in which we put emphasis on persistence and proper scaling of diagnostic attributes. Two of the paradigms of ecology we identify (Fig. 7.7), Things and Then, rely strongly on the conceptualization of ecological entities. The third one, Now, combines all three but Then, and even the Stuff paradigm would be severely impaired without recognition of entities such as ecosystems or their components. The ability to discern and repre-

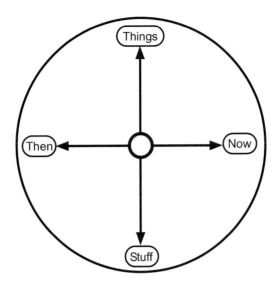

Figure 7.7   A target for integration in ecology. The two axes separating contrasting paradigms in ecology, Things versus Stuff and Then versus Now, may be considered to intersect and to indicate the aim of integration in ecology. An integrated ecology would bring the four perspectives together.

sent entities is crucial to the conceptual management of ecological complexity (Kolasa 2005). In view of the importance of the entities framework, a question emerges about the reasons for the limited progress in that area. While a number of ecologists recognize the need for consistent bounding of the systems they study and identify pitfalls. When one fails to do so, there is no widespread effort to adopt one or another framework. It is possible that one challenge facing a consistent and profitable use of the framework for ecological entities is that ecological entities come in varying degrees of discreteness. Some entities may be virtually invisible and yet have major consequences for the functioning of populations, species, or multispecies systems. For example, Californian song sparrows have song dialects that determine mating preferences. A bird imprinted on a particular song of local population prefers to mate with individuals singing the same song. Consequently, the coastal population effectively breaks into a series of regionalized subpopulations, each with its own population dynamics, parasite loads, predators, resources, and history. Each such subpopulation is best seen as a separate entity until the specific research question legitimizes lumping them together. Indeed, many other ecological entities such as breeding pairs, colonial organisms, island populations, lake communities, or symbiotic associations are more obvious, and ecologists do take methodological advantage of their discreteness.

From the perspective of theoretical integration and understanding, ecological entities offer a great promise. Entities, or ecological systems, engage in self-maintenance (Now paradigm), have history (Then paradigm), are Things, and rely on Stuff. In short, ecological entities may help coalesce all four paradigms under a unitary framework.

V. Theory as a Constraint on Integration across Paradigms: New Fundamental Questions

Integration and, therefore, fundamental questions, can be thought of as occurring on several hierarchical levels. Understanding, as we have defined it (Chapter 2), is itself an integration of

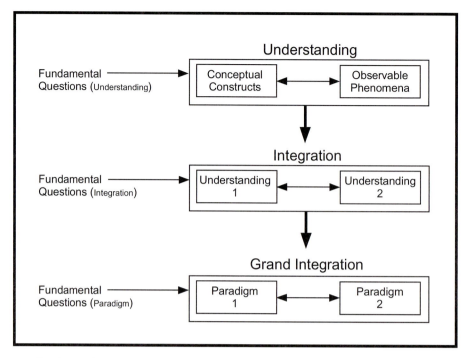

Figure 7.8 Fundamental questions can be focused on (1) changing understanding within a domain via a dialogue between conceptual constructs and observable phenomena, (2) changes in understanding that integrate across different domains, or (3) changes in paradigms. Moving from the level of within-domain to between-domain to between-paradigm parallels increasingly grand integrations.

conceptual constructs with observed phenomena. On the next higher level, integration can link two or more theories (Fig. 7.8). On the highest level, integration can link contrasting paradigms. The nature and status of theory are relevant to integration at each level; fundamental questions can be posed to stimulate synthesis on each of the levels (Fig. 7.8). Considering how theory can affect integration may point to fundamental questions, as well as caveats, for synthesis. First, integration across paradigms may be limited if theory in one of the areas is poorly developed. In general terms, theories in different ontogenic states (Chapter 4) may be difficult to integrate as wholes. However, in extractive integration (Box 7.5) across paradigms, different ontogenic states may be less problematical. Second, integration may be constrained by the failure of two theories to refer to a common currency. This is especially a problem in ecology, in which the focal units referred to by contrasting theories tend to hew close to what are often presumed to

---

BOX 7.5 Modes of Synthesis

Synthesis. Combination of two or more distinct theories as wholes
Extractive integration. Combination of parts of two or more theories into a new theory
Unification. Combination of the domains of two or more apparently disparate theories into a higher level theory; a special case of synthesis

Table 7-1 A Matrix of Paradigm Interactions Showing Status and Some Ecological and Related Disciplines Representing the Intersections

	Stuff	Things	Then	Now
Stuff	—	• To be developed	• Geology • Prebiotic chemistry • Pedology	• Ecosystem theory
Things		—	• Paleoecology • Paleontology	• Population ecology • Community ecology
Then			—	• Evolution
Now				—

be concrete units found in nature. For example, an ecologist studying a plant community undergoing change in response to global change might forget that what is actually being studied is a model of the community linked to a model of the climate. The material world is always seen through the lens of theory and its component models. A possible way to overcome the problem of inappropriate concreteness is to attempt greater abstraction within paradigms with an eye toward unification based on the new abstractions. Because so many ecologists choose their careers from a base of hands-on natural history, such a leap into abstraction may be difficult. However, data on the material world are amazingly abstract (Allen and Star 1982).

Even accepting the initial entry to ecology via apparently concrete units of nature, some likely fundamental questions for cross-paradigm integration can be suggested. First, ecologists may ask (cf. Lawton and Jones 1993), "How are the distribution and abundance of organisms influenced by the fluxes of matter and energy in the natural world and vice versa?" As a corollary, "How do ecological entities above the level of the individual (e.g., communities, land-use classes, and so on) behave in the reciprocal influences of distribution and abundance on fluxes?" To answer such questions, models of entities that include their characteristics as individuals and as processors of fluxes need to be developed. The concept that individual organisms reside at nodal positions in the major ecological hierarchies that reflect the four disciplinary paradigms (Mac-Mahon et al. 1978) suggests that new abstractions emphasizing both individuality and processing may provide a basis for the new integration. It is, of course, possible to erect other ecological hierarchies (e.g., Scheiner et al. 1993) or use other sorts of models that capture this dual feature.

Another related fundamental question deals with mechanisms of constraint or enablement offered by the extremes of each of the two axes. For instance, how does processing of material and energy constrain the structure of entities? It has been noted, for example, that oligotrophic freshwater and marine ecosystems often have higher diversity and richness than eutrophic systems. Likewise, how does the structure of entities constrain processing and fluxes? One could ask, for example, if the leaf litter processing in Australian streams differs from that in North American streams as a result of the influence of the chemistry of eucalyptus leaves. Similar questions could be asked about history and contemporary interactions. Explicit attention to the contrasting extreme along either of the paradigm axes is the first step toward integration of those complementary viewpoints (Table 7-1).

VI. Theory as a Constraint on Integration across Paradigms

A comprehensive theory for ecology that proved to be truly integrative would encompass the four paradigms identified earlier. Although it is impossible to envision the content and structure of such a theory currently, it is clear that ways will have to be found to translate among all the perspectives. A first requirement is to recognize the heuristic and empirical legitimacy of all four paradigms. Second, it is necessary to recognize that most important ecological theories of the past and present exist within one, or perhaps two, of the four realms. Progress toward integration of all the paradigms will be enhanced by a focus on theories and empirical studies that highlight neglected combinations of the two axes (Table 7-1).

Note that the four paradigms, because they involve tacit assumptions, can be dealt with objectively. Like any aspect of the dialogue between conceptual constructs and observable phenomena, paradigms are subject to analysis, if their assumptions can be exposed. Just as integration involves the objective addition of theories or a combination of components of different theories into some new area of understanding, so disciplinary paradigms can objectively contribute to integration, *provided the basic assumptions and contexts can be stated explicitly and in operational terms*, rather than remaining tacit and in the background. As in the case of changes in theories, fundamental questions for integration may be focused on the establishment, refinement, rejection, replacement, or extension of a disciplinary paradigm. For a revolution to occur, an alternative paradigm must be available, however. Disciplinary paradigms, as well as those that apply to ecology as a whole, are complex constructs embracing a variety of theories. Consequently, dealing with them will require extreme care and ability to discern, in concrete and operational ways, disciplinary assumptions that are hardly ever recognized.

Because disciplinary paradigms or ecological paradigms as a whole are complex and often tacit, the advice that graduate students attempt to attack a paradigm in their thesis research (Janovy 1985) may be problematical. Certainly, the smaller the scope of a paradigm, the closer it comes to equivalence with the broad concept of theory we have summarized (Chapter 3). Because theories should be well delimited, their parts should be clearly stated, and their internal linkages should be clear, it is good advice for graduate students to identify the relevant theory, state its components, identify the relationship of their research to the relevant components of the theory, and put their results in terms of the dialogue between conceptual constructs and observable phenomena. In other words, students should be able to answer such questions as these: What, explicitly, is the theory you are dealing with? What are its components, and which ones are missing or poorly formed? What are the roles of generalization, causal explanation, and testing in answering the questions posed by the research? What is to be tested, and what are the specific hypotheses? Which of the tests use confirmation, and which ones use falsification? What is to be explained, and what are the mutually reinforcing threads of the explanation? What alternative or complementary models are available for explanation? What facts are generalized, or what domain of generalizations is altered by the research? What facts are contributed to existing generalizations? What new generalizations are suggested? Is a theory threatened by the observations and tests? Positive or negative instances of these issues are valuable contributions, depending on the empirical status of the area. Such questions are manageable within the scope of graduate (and most other) research. Theory and the identification of fundamental questions are tools that can be used effectively in that context. Trying to identify and work with theory and fundamental questions that bridge the existing disciplinary paradigms in ecology may be a more productive avenue for graduate research than a threat to an ecological or disciplinary paradigm. Indeed, any research that exposes unstated background assumptions of disciplinary or ecological paradigms makes those assumptions and the connections they imply susceptible to objective interrogation by the entire scientific community.

VII. Conclusions and Prospects

Integration may occur across different domains of ecological understanding, by combining two complete theories, perhaps in a hierarchical fashion, to construct a new theory of broader scope. Alternatively, relevant elements can be taken from any number of different theories to construct a new theory.

The concept of paradigm has three relevant components in application to ecology: a world-view, a belief system, and a set of problem solutions. Unfortunately, the existence of a belief system as a component of paradigms can be incorrectly interpreted to indicate that science is subjective. In this chapter, we have focused first on paradigms as viewpoints. Paradigms can apply to science as a whole, to a specific area of science, or to disciplines within a broad area of science. The disciplinary paradigms in ecology are those that focus on the extremes of two axes: individuals versus fluxes and instantaneous versus historical coverage. These two axes define an ideal focal point for ecological integration that would incorporate all four perspectives, as necessary. There are potentially fundamental questions to stimulate integration across disciplinary paradigms, analogous to those for stimulating integration of different theories within a paradigm. All paradigms must be represented by clear, complete theories before they can be integrated.

To this point, we have explored the usually recognized objective aspects of science. Even the aspects of paradigms we have presented are those that essentially can be captured in the largest theories of a science. However, the concept of paradigm as advanced by Kuhn also includes the belief system of a science or subdiscipline. A belief system may affect integration as well, either positively or negatively. The next chapter explores how the belief system either constrains or enhances integration and how the community of scientists can operate objectively in theory change and integration.

Part IV

Theory and Its Environment

8

Constraint and Objectivity in Ecological Integration

"The prosaic, passionless formulas of scientific reporting serve an important purpose:
They maintain science as a communal enterprise, free of the prejudice of nationality,
race, religion, and personality that have plagued many human enterprises."
Raymo 1991:110–111

I. Overview

Paradigms exist at the level of disciplines within ecology, the entire science of ecology, and the entirety of contemporary science. To pursue radical integration that joins discrete paradigms, it is necessary to determine how both paradigms and theory constrain integration and how those constraints may be overcome. Before we examine how the larger paradigms within ecology impact understanding and the potential for objectivity, we must contrast subjective and objective approaches to science, from the perspective of practicing scientists. Knowing how changes result from integration in an open system of understanding leads to an appreciation of the "transforming interrogation" (Longino 1990) that the scientific community conducts on theory, phenomena, and their interaction. This is a process that upholds objectivity. We close the chapter by examining the relationship of understanding, culture, and paradigms. Specifically, we focus on the interaction between the objective changes in understanding, the cultural influences on the scientific community, and the role of those paradigms that affect the entire science of ecology.

II. Sociological Constraints on Integration

Two kinds of constraints inject bias into science. Both can be countered, or their significance in a particular situation can be analyzed, but only if they are recognized. The first kind of constraint is the social component of a science. This kind of constraint is responsible for a bias that operates *within* the scientific community. For the sake of brevity, we will call the internal social constraints within science "sociological," even though a sociologist might not find this an entirely suitable use of the word. The second source of bias originates in society at large. We will call constraints originating from the larger society "societal." Again, we have perhaps chosen an ugly word, but such unfamiliar terms help highlight the negative connotations of subjectivity that the

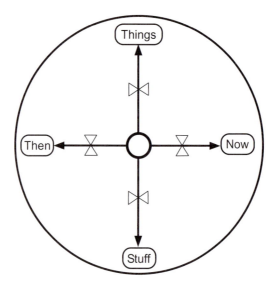

Figure 8.1   Control gates on integration. Integration can be limited by differences in theories, paradigms, and sociological factors that stand between the contrasting foci in ecology. The controls are indicated by the valves
or "bowties" along the paradigm axes.

constraint involves. Both kinds of bias can influence integration across paradigms. In fact, these biases compose a kind of constraint on integration that originates outside the realm of the disciplinary paradigms discussed in Chapter 7. The external societal and internal sociological structure of a science are constraints that act as control gates on the activities of scientists. They can restrict intellectual exchange along the axes of Things-Stuff and Then-Now, which describe the disciplinary paradigms of ecology (Fig. 8.1).

Within science, the sociological constraints on integration are of three overlapping kinds (Fig. 8.2). One is scholastic, one methodological, and the last is personal.

A. Scholasticism

Scholasticism reflects the training of scientists within a "school" or laboratory (McIntosh 1985). Often scholasticism results in professor/student lineages that share subject matter, approach, publication outlets, philosophy, and desired rewards. Lineages can also form a personnel network, supplying new students and exchanging postdoctoral training. Although it is entirely possible that a school could adopt an integrative approach and the necessary reward system and phi-losophy to support it (Pickett et al. 1999), the need to differentiate a school may, unfortunately, be more easily served by claiming a distinct paradigm and focusing on a specific theoretical area and set of tools, as well as communicating primarily with others within the same paradigm. Scholasticism may therefore often act as a brake on integration.

An example of scholasticism is found in the early history of American plant ecology, in which the mechanistic- and community-oriented practitioners traced their academic relationships to William S. Cooper at the University of Minnesota. Indeed, all the dominant textbooks of the second generation of American plant ecology emerged from this school, and the major ecologi-cal journals were edited by people in the lineage. Their focus was on the structure of the plant community and the controls on the distribution of plants. These areas can be said to represent

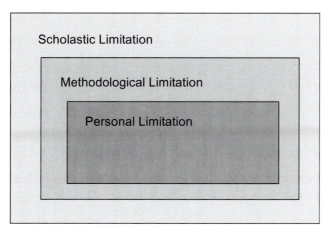

Figure 8.2   Concentric limitations on integration in science. Integration is limited by personal talents and training, methods available, and scholastic lineages and associated rewards.

a vertical integration, but still the scope was specific to plants and the abiotic environment. The focus of this school becomes even sharper when contrasted with the European and Russian traditions of community plant ecology, which focused on classification of vegetation units on the landscape and their relationship to coarse-scale soil patterns (McIntosh 1985). The American school, in contrast, focused on plant-environment relationships in the lineage descended from Cooper. This school also emphasized the role of plant-environment relationships in the successional dynamics of vegetation.

A second example of a school is the conscious effort by Eugene P. Odum to legitimize ecology as a discipline distinct from the then dominant biological fields of botany and zoology which, if they claimed ecology at all, were perceived to do so as a footnote. He emphasized the ecosystem as a separate level of organization worthy of study, sought its emergent properties, and published the first textbook in the area (Odum 1953). Here is an example of a school of thought that was founded to accomplish a new disciplinary synthesis (Golley 1993, Hagen 1992).

Of course, within each of these schools, there were important integrations, but the four large disciplinary paradigms we have identified (Chapter 7) provided separate homes for these schools. Integration among schools was not forthcoming for a long time, and the limit was sociological rather than determined by the subject matter. Of course, to be fair, these and other schools that were responsible for founding or consolidating the young science of ecology (McIntosh 1985) had to focus relatively specifically to make progress and to claim academic territory. However, building on their foundation, contemporary ecologists now can see the intellectual landscape in which cross-paradigm bridge building is such a large opportunity. The shape of the ecological landscape would not now be clear without the efforts of the founding schools. However, continued narrow fidelity to a scholastic tradition in ecology would inhibit the largest integrations that the field can now support. Perhaps with the involvement of more institutions in the training of ecologists, the ease of travel, the career mobility of professionals, and the proliferation of meetings and workshops, scholasticism may become less of an impediment to integration. One especially notable tool ecologists have to promote integration and synthesis is the National Center for Ecological Analysis and Synthesis, in Santa Barbara, California. However, McIntosh (1985) noted that scholasticism can operate through "invisible colleges," which we can liken to a virtual intellectual reality. The ability of scholasticism to be camouflaged as invisible colleges and to continue to apply friction that slows synthesis should be guarded against.

B. Methodology

A related sociological phenomenon that can inhibit progress is the discipline-wide training bias toward a particular kind of system or a method and the specific set of skills associated with it. Ability to measure nutrients, population density, spatial patterns, or many other attributes of ecological systems is indispensable in conducting effective research. However, the habit of reflecting on the direction of this research is not a systematic part of ecologists' education, and the reflection itself is largely left to those further advanced in their careers and willing to make ad hoc forays into the intangibles. Consequently, for many years a great number of ecologists may not have the opportunity to contribute to the growth of ideas, even if they may have valuable contributions to make.

Methodological limits to integration take many forms. Whether the studies should be primarily experimental or comparative is one such matter of choice (Grime 1979, Tilman 1989). So too is the decision to search for single causes or to evaluate spectra of causes. The study of population regulation and cycling has exemplified this contrast (Gaines et al. 1991, Lidicker 1988). One group argues that the strategy of testing single factors at a time has resulted in discarding factors that do, in fact, affect the population size of mammals, so a multifactorial approach is necessary to understand population regulation (Lidicker 1988, 1991). In contrast, others suggest that only by attacking factors in isolation can the problem be convincingly solved (Gaines et al. 1991).

This debate about multiple versus simple causality continues with little convergence of positions. Naeem (2002) posited that ecology is changing its paradigm from the general view that diversity is something to be accounted for to a position where diversity becomes a player in producing ecological phenomena. In this context he characterized the tension between those who seek to explain nature by studying its parts and those who seek to explain nature by studying whole-system behavior. However, although we reviewed earlier (Chapters 1 and 4) how both perspectives interact synergistically in generating understanding, ecologists have yet to embrace this fact and adopt it as a constructive strategy. This tension may be an unnecessary hindrance that emerges from extrascientific prejudice rather than from scientific imperative. Still another example of methodological debate, with implications for research focus and funding, involves the role of model systems. Model systems have three useful features — tractability, generality, and realism (Levins 1966) — which enable future experiments to build on previous results. Ecologists have adopted many models that meet some but not all of these requirements. Indeed, not all three can be maximized simultaneously. For example, laboratory-assembled communities of protozoa enable quick, precise, and highly replicated experiments, which represents high tractability. However, this strategy has been criticized for its artificiality and, hence, potentially low generality. Whole-ecosystem experiments represent the opposite extreme. This strategy achieves high realism by manipulating entire natural communities whose large size usually necessitates poorly replicated and mechanistically simple studies. Thus, whole ecosystem experiments may suffer from potentially low tractability. There has been heated debate about the apparent merits of replication versus realism (Schindler 1998) in ecology. Natural microcosms potentially offer a way to circumvent this trade-off between artificiality and tractability (Srivastava et al. 2004). It should be clear by now that the most promising strategy will likely involve an integrated use of all three approaches, and the least promising strategy will emerge if one approach suppresses the others.

Another contrast in methodology is the use of a hierarchical approach or a predominantly single-level approach, as seen in the contrast between the old individualistic and community unit debate (McIntosh 1985). In this context, the community represents a level of organization. The issue is whether communities are characterized by species that have discrete and coincident distributions or are characterized by species distributions that are continuous along gradients, as has been argued since Gleason (1917) first criticized Clements (1916). Part of the resolution is to determine whether the species *behaviors* are individualistic or whether the resulting distributional *patterns* are (Shipley and Keddy 1987). The argument has for most of its history focused on the patterns on the community level of organization alone and has neglected the hierarchical division of pattern into contributory processes on lower levels of organization, such as the population (Parker 2004).

As another example, we can cite the renewed research and debates on the diversity-stability issue. Whether the concern is about species richness or functional diversity, models as well as laboratory and field experiments involve mixes of species without any consideration of hierarchy of any kind, whether organizational, spatial, or temporal (e.g., McCann 2000, Tilman 1999). Given that ecologists have identified the need for inclusion of scale and hierarchy as an important tactic for dealing with complex issues, the asymmetric approach to the diversity-stability question may hinder integrative progress.

Finally, a research approach may be broad or specific. Such a contrast is found in studies of the ecological anthropology of a particular people or village (Boyden 1987), compared with the study of the resource interactions and other influences, whether they be local or not (Vayda 1983). Vayda has provided an intriguing term for the process of following the implications of a particular ecological relationship into new contexts of interaction to explain the original relationship. He calls such a strategy "progressive contextualization" and suggests that its hallmarks include an ability to look beyond preconceived boundaries and limited rosters of controls. Of course, certain subject matters may be better suited for combining these contrasting approaches in different proportions, but an approach is so often associated with particular laboratories or schools that some sociological inertia seems to be at work as well.

Carpenter (2003) has presented a broad view of ecological methodology. Using the metaphor of a four-legged table, he suggests that ecology is functional because it too has four legs: experimentation, observation, modeling, and long-term studies. Many ecological studies, and certainly the discipline as a whole, find success in using the entire suite of methods represented by the four legs of Carpenter's table. Integration of methods is thus one of the strategies for success in ecology. The breadth of methods available to ecologists and the need to meld models with data from various sources are also emphasized by Ford (2000) and by Hilborn and Mangel (1997).

C. Personality

The third aspect of sociological constraint within science is in the personality of individual researchers. Personality may determine what school, approach, or topic a person chooses. Alternatively, personality may also determine what school chooses a person! Examples of personality traits that may be differentially expressed in the choice of a school, topic, or approach include whether a person is (1) critical or constructive, (2) quantitative or qualitative, (3) abstract or concrete, (4) statistical or deterministic, (5) experimental or analytical, (6) practical or conceptual, (7) technical or nontechnical, (8) mathematical or verbal, and so on. This is certainly neither a complete nor a mutually exclusive enumeration of personal proclivities. It is meant to illustrate the range of personality traits that may influence the science a person does. The sociological and personal features can affect the choice of paradigm a scientist works in and also whether a scientist attempts integration among paradigms.

III. Societal Constraints on Integration

The second major source of biases that may affect the conduct of science in general and of integration in particular is the larger society and social context, in contrast to the sociology of science itself discussed in the previous section. Societal biases can affect the choice of paradigm, approach, and method. By virtue of space and training, we cannot attempt to analyze these sources here, but we merely mention them for completeness. What sort of science one does may well relate to culture and ethnicity, class and gender, and the rewards society offers for conducting research. The societal rewards affecting science deserve some additional attention. They include funding for research and personal remuneration, but personal notoriety can also be a factor. The problem of funding is especially important in limiting integration because funding decisions are usually made within recognized disciplinary boundaries (Pickett et al. 1999). The institutional nature of funding reflects the disciplinary paradigms in ecology and may therefore inhibit integration.

The societal biases may affect research because they inject hidden assumptions into the structure and conduct of science. Such hidden assumptions from outside science can determine what counts as good research, what sort of models of a system or problem are valued, and what sorts of answers are acceptable. An example might be scientists from a culture that values individual action proposing a model of social organization that reflects a linear dominance hierarchy, versus scientists from cultures that value communal action proposing a cooperative model for the same kind of animal. Similarly, the rise of feminism has encouraged some kinds of research, say, into women's health issues, that previously had been neglected (Schiebinger 1999). Sophisticated analyses of societal constraints in science are provided by Longino (1990).

The societal constraints only remain problematical, however, if they persist as invisible shared assumptions. In a hidden form, such assumptions constitute the belief component of a Kuhnian paradigm. We purposely left that aspect of paradigm unexamined until now. Earlier, in Chapter 7, we concentrated on the more traditionally recognized, objective components of the disciplinary paradigms in ecology. To the extent that outside assumptions can be made visible or explicit, they can be made the *object* of analysis by the scientific community. Objectification requires an alternative view for comparison that different beliefs view as theoretical assumptions. Objectification of beliefs into explicit, articulated assumptions brings them into the world of science. To see better how to deal constructively with the paradigms that affect an entire scientific field, the nature of objectivity in science must be addressed.

IV. Scientific Objectivity and Changes in Paradigm

A full awareness of the nature of scientific objectivity may help overcome the problems of societal and sociological constraint on integration. This analysis of objectivity also will help flesh out other important facets of scientific understanding. Analysis of scientific objectivity still suffers from the persistent positivist model of science (Chapter 1). If, as positivists assumed, science is conducted by individuals, then there must be a way to prevent their subjective biases from intentionally or unintentionally affecting the conclusions they draw. The positivists attempted to establish rules of inference, suggested by the nonempirical systems of geometry and logic, to prevent an individual from introducing incorrect statements into his or her theoretical system of statements. We explained earlier that these nonempirical systems of understanding can be considered closed, since they do not depend on contact with the external world for standards of proof. Popper, following this positivist tradition (Hacking 1983), established the falsificationist doctrine to discriminate meaningful scientific statements from nonscientific statements. In essence, the falsificationist view states that if one could not potentially falsify a statement, then it was not part of science. The continued analysis in philosophy and history of science ultimately led to the demise of the positivist philosophy, including its strict falsificationist component (Boyd 1991).

If logical positivism, with its model of rules designed for closed systems of understanding, could not unambiguously tell individual scientists how to behave objectively, then some alternative was needed. Kuhn (1962, 1970) provided an alternative, which is paralleled in an apparently extremely subjective form by Feyerabend (1975). In the alternative view, advance in science must be driven by consolidation of a worldview and belief system within a scientific community and by the overthrow of one worldview by another contrasting one after a crisis. This is the view of the history of science as a series of revolutions overturning worldviews that are set for some period of time.

The situation may appear desperate to many scientists and to observers in the public. Either science, which is empirical, must follow strictly logical rules that are more appropriate for non-empirical, closed systems of understanding, or it must follow a subjective community-level anarchy. Fortunately, both of these troublesome conclusions share a common assumption (Longino 1990). The positivist model and its alternative in Kuhnian revolutions and the like all assume that science is conducted by unitary entities. This entity is either an individual scientist or a monadic community of scientists.

The alternative approach to scientific objectivity considers science to be conducted by a diverse, open community of practitioners (Longino 1990). This view is one that seems to us as practicing scientists to reflect well the reality of science. It is apparently a new approach in the philosophy of science, with only a short history. Under this new view, called contextual empiricism (Longino 1990), there are two components of scientific objectivity. One resides in the science itself and reflects the constitutive values of science. In essence, the ingredients of the constitutive aspect of objectivity were presented in Chapters 2 and 3. Objectivity originates within science based on an *open* system of understanding, which has specific components that can be stated and can be compared to observable phenomena in several ways. Scientific understanding is said to be open because it engages the material world and can change as a result. This open approach to understanding therefore embraces multiple modes of relating reality with theory and employs several cross-checks on the goodness and appropriateness of the fit of conceptual constructs to observable phenomena (Lloyd 1987, 1988, Salmon 1984). There are even rules for choosing one theory over another, when two theories are both successful in establishing a general explanation. Preferred theories will be those that are accurate, consistent internally and with other accepted theories, broad in scope, simple, and fruitful of new research (Kuhn 1977).

The objectivity of science has another component that originates in the context of interaction within a diverse scientific community. To ensure that the dialogue is as free of bias as possible, science is the product of a community that uses an open, nonauthoritarian system of understanding. *Nulius in verba*, the motto of the Royal Society of London, dating from 1660, embodies these two traits. It translates roughly as "No one's word is final." The openness of the system of understanding invites novel empirical insights into the dialogue between conceptual constructs and observable phenomena and allows the shape and content of the resulting state of understanding to change. Science results in understanding and explanations that are inherently correctable.

The reduction of bias is also enhanced by the participation of people with different intellectual proclivities. Some will pursue synthesis and idea generation, whereas others will criticize the syntheses and new ideas; some will compare and extract pattern, whereas others will experimentally explore mechanisms; some will abstract or conceptualize, whereas others will probe systems on the ground in all their wonderful messiness. The exchange within the community will be conducted by people with different, contrasting technical skills. There will be instrument freaks, computer wizards, chemical demons, modeling fiends, masterful experimental architects, abstract thinkers, keen-eyed natural historians, profound scholars, nitpickers, and dreamers, among others. In fact, one person may combine several of these disparate traits or express different traits or combinations of traits over time. Finally, the scientific community will be composed of people of two sexes and several sexual orientations; people from different social classes and positions of social privilege; people from different racial, ethnic, or cultural backgrounds; and people with different political bents and (often tacit?) philosophies.

The social biases are purged from science when people from various backgrounds identify some previously unstated or unsuspected background assumption that has become part of a paradigm, and therefore open the assumption to analysis. The identification of unspoken background assumptions is critical, because these assumptions often underwrite some particular theoretical structure, methodological approach, or empirical interpretation. Whether the social biases act as assumptions that affect the conclusions of a study or discipline, whether they are empirically justifiable or not, and whether they have possible replacements are all issues that can be addressed by the scrutiny of a diverse community. The examination of background assumptions by a culturally diverse community is an important part of the "transformative interrogation" of theory and observation that constitutes science (Longino 1990). Alternative, explicit theories and new themes in the dialogue with observable phenomena can be established as a result of discovering contrasting cultural biases that had acted as background assumptions for science. The theories, the observations, and the multifaceted dialogue between them, conducted by diverse practitioners, generate objective, scientific understanding that is as free of bias as possible. In summary, we may say that objectivity of science is produced by a mutual cancellation of subjectivities. Such cancellation is possible through diversity of personalities, schools, and approaches. Bias cannot be avoided, but it can be countered in an open community that respects diversity. Even the intentional introduction of bias for personal or political gain can eventually be exposed by this grand intellectual strategy.

Thinking about scientific objectivity as a process driven by a heterogeneous community of practitioners suggests a useful statement of the nature of science to counter the one that is a holdover from the failed philosophy of positivism. If positivism has warranted too narrow a definition of science, one that is troublesome in courts of law and public discourse, perhaps a new one that is not positivist will be more useful. Of course, a new slogan for science will necessarily be more complex than one underwritten by positivism, because positivism only valued a small part of the scientific enterprise: its strictly logical structure. Therefore, combining the conclusions of Longino (1990) with concepts discussed earlier in this book, we offer a definition:

> Science is the process of transformative interrogation by a diverse community of investigators that results in an open system of understanding focusing on some structure or process in the material world.

Hence, we see science as a community endeavor (Fig. 8.3) that develops and transforms empirical and conceptual constructs as a means to understand nature. Scientists accept seven major characteristics and goals for their activities:

1. Their vocation is preeminently a community effort.
2. They can be wrong.
3. They must be able to change their minds.
4. Their community is obliged to expose assumptions, whether they arise from within science, self, or society.
5. The implications of assumptions must be rigorously explored to the fullest extent via the mechanism of theory.
6. Inference from these assumptions may be logical, statistical, causal, or some combination of these.
7. Understanding is dynamic and can grow or change based on the dialogue between concept and observation.

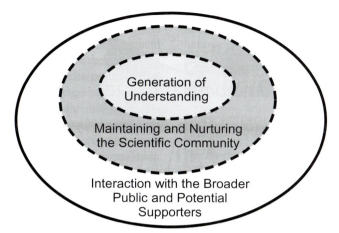

Figure 8.3   Science is a complex, layered activity. At the core is the activity of understanding the pattern and process, including change, in the material world. This core activity is enveloped in the activity of a community, which must be constructed and nurtured. Finally, the scientific community and its work are placed within a social context, to which it owes its existence and from which it seeks its support. In Chapter 9, more will be said about the outer, public layer.

These seven characteristics show science to be a much more complex endeavor than the old view of positivism or even the hopeful but limited descriptions found in textbooks or public accounts of science. Most scientists probably already know this at an operational level that guides their day-to-day activities. However, the complex, interactive, and transformative nature of the process and dynamics of science is hardly ever articulated by scientists. We hope that this characterization can help clear up apparent contradictions between what scientists do and feel to be correct procedure and the simple and often incorrect admonitions of narrow and out-of-date descriptions by some philosophers of science. Furthermore, this characterization puts to rest superficial and pointless debates about soft and hard science. It is not the properties of subject matter, or the toolbox of methods, or the formalisms that define science but rather a commitment to the systematic and objective effort to query nature, and that determines the standing of a discipline within the family of scientific disciplines. In this sense, ecology, physics, geography, and animal behavior are all equal sciences.

We have now examined integration and have indicated how paradigm choices and their social contexts might limit integration across disciplines within ecology. Several other paradigms affect the entirety of ecological science. We turn to them now. Just as societal and social constraints may restrict the integration between the disciplinary paradigms of Things-Stuff and Then-Now (Fig. 8.2), so too can the major paradigms affecting the whole of ecological science constrain integration. The next section presents examples of the role of large scientific paradigms and their integration in ecology.

V.   Integration and Paradigms Affecting the Whole of Ecology

We have previously clarified the concept of paradigm to indicate that paradigms of various degrees of breadth or inclusiveness were possible (Box 7.1) — that is, there is a hierarchy of paradigms. There is a paradigm for contemporary science as a whole, within which are nested the paradigms of specific sciences such as physics, chemistry, ecology, and social science. Within each of these broad disciplinary paradigms, there are also more focused, specific paradigms that cover different topics within each science. Preceding sections of this chapter have focused on the disciplinary paradigms that appear within ecology. Here we discuss paradigms that run through the whole science of ecology. These paradigms are strongly influenced by cultural beliefs as well as scientific viewpoints. It is important to discuss these paradigms because they are resolving into a new configuration. These ecological paradigms affect the dialogue between observable phenomena and conceptual constructs in all of the more specialized disciplinary paradigms within ecology. The resolution of new ecological paradigms is an example of the working of the model of objectivity via transformative interrogation by a diverse community.

We will examine three overarching ecological paradigms. These paradigms appear as contrasts between (1) the equilibrium and the nonequilibrium worldview, (2) the unilateral control and reciprocal control view, and (3) the simple and multiple causation view. We will devote most of our attention to the first contrast, which is in a more advanced state of resolution.

A. Equilibrium and Nonequilibrium

The equilibrium paradigm is one of the oldest and most pervasive ideas in ecology. It is built on worldviews that have been deeply rooted in Western cultures since at least the ancient Greeks (Botkin 1990). Tenets of the equilibrium paradigm consider ecological systems (1) to be essentially closed, (2) to be self-regulating, (3) to possess a stable point or stable cycle equilibria, (4) to have deterministic dynamics, (5) to be virtually free of disturbance, and (6) to be independent of human influences (Botkin 1990, Pickett et al. 1992, Simberloff 1980). A large disciplinary paradigm changes only over long periods. The period of time required to replace the equilibrium paradigm is essentially most of the first 90 years of the history of ecology as a self-conscious science in North America. Two changes in ecology led to the abandonment of the equilibrium paradigm. First, empirical evidence accumulated that made the assumptions visible and ultimately successfully challenged them. Second, ecologists have been able to view their systems at different, often larger, scales than were percievable through much of the first half of the discipline's history. We next present some of the kinds of evidence brought to bear on each point of the classical paradigm.

1. Material Openness

As functional attributes and processes within ecological systems became increasingly better examined, many ecological systems were found to be materially open. The laws of thermodynamics dictate that systems be open to the flux of energy, and ecologists have always accounted for this. However, fluxes of materials, organisms, and information across system boundaries are now known to be common. For example, edges are considered to be important in wildlife biology and in landscape ecology (Cadenasso et al. 2003, Holland et al. 1991). However, even subjects that have ignored boundaries and openness now recognize their importance. For example, an understanding of successional processes in old fields required knowledge of adjacent vegetation (Armesto et al. 1991, Meiners and Pickett 1999), and stream and lake ecologies have benefited from a watershed approach that sees those water bodies as linked with adjacent sites (Likens 1992).

2. External Regulation

The self-regulatory capacity of many systems was questioned once their long-term dynamics became available, and the rarity of numerical constancy over time became apparent (Strong et al. 1984). A stable point equilibrium failed to be a useful descriptor or predictor of systems on many critical scales of interest (Botkin and Sobel 1975). For instance, the postglacial vegetation shifts in temperate areas showed that different species occupied geographically distinct refuges during full glacial times and followed different reinvasion routes and rates (Davis 1983). Thus, the eastern deciduous forest of North America did not respond as an integrated assemblage to glacial advance and retreat. There was not a unified, self-regulated system of species that came and went in the same way. Similarly, even many tropical forests have undergone large geographic shifts as a result of changes in rainfall patterns during and after the Pleistocene. While so much of the world's water was locked up in the high latitude glaciers, there was less rainfall at lower latitudes, leading to migrations of species.

3. Point Equilibrium

An equilibrium point is rarely encountered in ecological systems. In instances where equilibrium appears to be a useful descriptor, it is an equilibrium *distribution* of patch types or some other attribute, rather than a persistent point equilibrium (Bormann and Likens 1979). In some vegetation studies, the actual spatial scale of the system is insufficient for even an equilibrium distribution of patches to arise, even if the extent of the system is large and contains many patches (Romme 1982).

4. Deterministic Dynamics

Determinism would require a strictly repeatable sequence of events. However, such lock-step change has rarely appeared, despite considerable efforts to detect it (Pool 1989). Consequently, successional pathways are now considered to be site and situation dependent (Pickett and Cadenasso 2005). Similarly, in animal community organization, priority effects are well recognized (Morin 1984), with the order of arrival affecting the outcomes of competitive or predator/prey interactions. The lack of rigid determinism is encapsulated in the term "contingency," which means that the current state of an ecological or evolutionary system is dependent on the specific conditions that may have occurred from time to time in the past or on the order of events that affected the trajectory (Cooper 2003, Gould 1989, Lawton 1999).

5. Rarity of Disturbance

Natural disturbance has been well documented in various systems (Walker 1999, White 1979). It is now apparent that, far from being an extraneous or exceptional event, many systems are in fact highly dependent or contingent on disturbance and its spatial and temporal distribution (Pickett and White 1985, Reice 1994). Even large, infrequent disturbances have been significant in system organization (Dale et al. 1999). Floods, fires, windstorms, land slides, herbivore outbreaks, and many other episodic events have periodically altered the structure or composition of ecological systems in ways that have persistent or large transitory effects. Stress, an environmental impact that at least initially alters the function or some process within a system, can have analogous effects. Sometimes those effects operate through subsequent changes in system structure and so would be recognized as disturbance, and sometimes they remain in the functional realm.

6. The Role of Humans

Finally, the failure to include human effects in natural systems distorts our understanding of ecological systems (Boyden 1993, McDonnell and Pickett 1993, Turner et al. 1990). Even in systems that had been thought to be pristine, indirect or distant effects of industrial societies are present. Nonindustrial societies, in the past and the present, including some having relatively low population densities, have had far-reaching or long-lasting influences on ecological systems (Burke 1985, Padoch 1993). A result of this shift is that ecologists are now becoming more engaged in research on inhabited systems, even including work on urban areas (Grimm et al. 2000, Pickett et al. 2001). Of course, the expanding density, extent, and global reach of urban areas motivate some of this research. During the decade begun in 2000, the global population will experience an urban tipping point, at which time more than half the Earth's population will live in urban areas. This remarkable shift, especially in South America, Africa, and Asia, will bring about massive changes in biota, land cover, climate, and topography (Vitousek 1994). Understanding the ecology of urban areas and the environments they influence is becoming a major need in ecology (Alberti et al. 2003, Grimm et al. 2000, Pickett et al. 2001).

In addition to the sorts of data mentioned for each point of the paradigm presented here, there is another kind of change that has motivated the paradigm changes. A literal shift of perspective has diluted the hegemony of the equilibrium paradigm. Ecologists have begun to examine their systems at coarser spatial scales. The shift in spatial scale has permitted ecologists to see the fluxes between systems, to learn the exogenous origins of the regulating factors in many systems, to assess equilibrium distribution of patch types in landscapes, and to appreciate the common-ness of disturbance (Weatherhead 1986). The shift in spatial scale is analogous to the temporal shift in scale discussed earlier, which was so influential in appreciating the long-term trends that contributed to the demise of the classical paradigm.

As a result of the empirical insights and scale shifts, ecology has developed a new paradigm, or new ecology (Cooper and Ruse 2003). We call it the nonequilibrium paradigm (Pickett et al. 1992), not to suggest that equilibrium never appears in nature but that it need not appear at all scales or for all phenomena. Under the nonequilibrium paradigm, ecological systems can be thought to be open, to be regulated by factors internal and external to them, to lack a stable point equilibrium, to be nondeterministic, to incorporate disturbance, and to admit human influence (Botkin and Sobel 1975). If an equilibrium is to be found in certain ecological systems, it may only appear on certain specified time intervals and at certain coarse spatial scales (Clark 1991). Equilibrium can thus be seen as a special case in an array of possible conditions in which ecological systems present themselves to researchers. Whether equilibrium appears or not often depends on the scale at which the system is examined, on the relationship between the system size and rate of change, and on the size and rate at which nonequilibrium factors act.

Two caveats must be emphasized about the equilibrium to nonequilibrium paradigm shift. The shift has been gradual (or perhaps fitful) and has taken place over a long time. There were numerous early critics of the strict equilibrium viewpoint. Finally, the new nonequilibrium paradigm is inclusive. Like so many ecological concepts, a gradient of ideas is implied by the nonequilibrium paradigm. Some situations may yield behaviors close to those expected of equilibrium systems, but as conditions become gradually more like those of a pure nonequilibrium system, so too will the behavior of the system change. A conceptual gradient thus recognizes that systems can be at equilibrium or may have an equilibrium distribution of states, but that not all ecological systems necessarily have an equilibrium. The existence of a continuum between equilibrium to nonequilibrium behavior in ecological systems means that an equilibrium point or distribution may be a useful point of reference in theory or in models, but systems on the ground may behave far from equilibrium under the influence of certain current environmental factors or as a result of history.

The shift from the equilibrium to the nonequilibrium paradigm illustrates the role of cultural biases in science. The idea of the "balance of nature" is not a scientific concept. It is a long-held cultural myth or metaphor for how the world works, dating in the Western tradition from the time of the ancient Greeks (Egerton 1973). With its implication of harmonious working of nature, this idea may have predisposed scientists to accept the equilibrium paradigm and may have delayed recognizing the implications of data that contradicted the equilibrium paradigm (Botkin 1990). It may also contribute to the evident discomfort of some scientists (e.g., Higgins 1975) with the parallel paradigm shift in the science of geomorphology from the erosion cycle to process geomorphology.

The idea of the balance of nature has influenced — and continues to influence — what and how science is used by managers and planners. If scientists, who have the weight of empirical evidence to consider, still fall into equilibrium or balance-of-nature thinking, how much easier is it for nonscientists to do so? However, there is, simply, no single or persistent balance of nature (Botkin 1990, Pickett et al. 1992). It is time for a fair turnabout from science to society. Perhaps the scientific appreciation that a strict balance of nature is rare or unlikely means that the insight can be translated to society. Is there a metaphor based on contemporary scientific paradigms that might inform societal dialogue about the structure and function of the material world as ecology understands it? The metaphor of "flux of nature" is a candidate (Pickett et al. 1992). This metaphor appears better to reflect ecological reality and may be useful in public dialogue. It emphasizes the dynamism, uncertainty, and contingency of ecological interactions and structures.

The remaining two major paradigms we touch on only briefly to indicate how their influence can extend throughout ecology.

B. Unilateral versus Reciprocal Control

The most common background assumption about how ecological systems are controlled is that control resides in the abiotic environment and affects the biotic element of interest (Fig. 8.4). Thus, climate and soils are usually considered to be independent variables determining vegetation patterning and performance. The physical environment determines the autecological responses of individuals, which are constrained by their genetic and metabolic limits, of course. However, organisms can adjust elastically to short-term environmental changes through development,

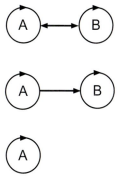

Figure 8.4   Reciprocal versus unidirectional causality. The middle diagram shows the usual ecological approach to problems — that is, assuming that phenomenon A causes phenomenon B. The lower diagram shows the discarded viewpoint, illustrated by the superorganism approach to communities, in which the causation of changes and structure of a system resides entirely within that system. The upper diagram indicates the multidirectional approach to causality that is increasingly being applied and explored in ecology. Nonlinearities, lags, and indirect effects add complexity to reciprocal relationships.

morphology, and physiology, and they can adjust plastically through long-term or slow evolutionary adaptations. Because organisms and the systems containing them can change so dramatically, the viewpoint of reciprocal control would seem to be the null model for interactions in ecology. Despite this, the Cartesian independent-dependent variable paradigm has prevailed in ecology. However, certain ecologists recognized quite early that reciprocal relationships could be significant in controlling the structure and behavior of the system of interest. Clements (1916) proposed a reaction by the plant community on the physical environment as a major mechanism structuring plant communities and driving successions. This viewpoint, which languished with the backlash against Clements's incorrect, nearly mystical holism of the superorganism, is being mechanistically revived in an appropriate form in plant ecology (Roberts 1987, Tilman 1988) and in general (Jones and Lawton 1995, Jones et al. 1994, 1997, Naeem 2002). The Lotka-Volterra models of predator/prey interaction emphasize reciprocal control and have been components of ecological theory for a long time, although they are often problematical in detail and application.

One shift in perspective about the locus of control has garnered public attention. The classical assumption about the chemistry of planet Earth considered it to be an independent variable influencing organisms and organism systems. However, the reciprocal relationships of the biota and abiota have been widely recognized as a result, for example, of the writings of James Lovelock (1979). Simply put, in its legitimate scientific form, his proposition is that geochemistry is influenced by the biota and that geochemistry in turn influences the biota. In other words, the Earth is an ecosystem. The reciprocal relationship is evidenced by the change in atmospheric chemistry over evolutionary time as oxygen-producing organisms came to prominence. This appropriate "biotification" of geochemistry has unfortunately entered the popular mind with all the cultural connotations of the balance of nature. Adding "bio" to geochemistry to produce "biogeochemistry" requires no such connotations and instead can rest on sound scientific concepts. These include the law of conservation of matter, the recognition of both positive and negative feedbacks, and an understanding of the nutrient and energetic limitation of organisms. The popular Gaia hypothesis seems to be an unnecessary extreme and a return to extremely metaphoric or "mythopoeic" thinking by some (Egerton 1993), although this metaphor is moving toward a more scientific form (Kirchner 2003).

There are, of course, some scales and processes for which control is indeed unilateral. The structure of a temperate stream community may be governed primarily by the velocity and trophic inputs from upstream or by the input of coarse woody debris from the watershed (Vannote et al. 1980). An additional kind of external and directional abiotic control appears in large physical disturbances. Many biotic assemblages are periodically limited by physical catastrophes, yet nonmystical reciprocal control is an important area for contemporary ecological exploration. A view that ecological systems, whether self-organizing or arbitrarily defined, are shaped by a mix of controls is best reflecting the current state of knowledge. This mix is projected as compound patterns that ecologists register and attempt to account for (Fig. 7.4). This paradigm contrast is still being resolved.

C. Simple versus Multiple Causality

A third major paradigm that affects the entirety of ecology is the contrast between two opposing modes of causality. One can search for a single cause for natural phenomena or for multiple, perhaps interacting, causes. This paradigm contrast has an anchor in the same Cartesian viewpoint that influenced the assumptions about the directionality of causality. The contrast between multiple and single causality is a persistent point of contention. The calls for using the method of strong inference (Platt 1964), in particular, have reinforced the nearly universal, superficially commonsense notion that an event has a single cause. There has been a long tradition of examining single causes in ecology. This is enshrined in the factor-by-factor accounts of classical texts (e.g., Andrewartha and Birch 1954, Daubenmire 1974, Krebs 1985) and appears as well in the sequential falsification of potential causes of, for example, population regulation, which often leaves one without a satisfactory causal explanation (Hilborn and Stearns 1982). It is effectively impossible to evaluate multiple causes using independent experiments aimed at examining interactions cause by cause. However, the physical magnitude and statistical complexity of multifactorial experiments sufficient to determine factor interaction and to interpret those interactions biologically are large (Gaines et al. 1991). Still, contemporary philosophical accounts recognize that the structure of theories and explanations incorporates networks and suites of causes (Miller 1987, Salmon 1984). Hilborn and Stearns (1982) examined the paradigm of multiple causation in ecology and proposed a multicausal methodology as the antidote. A more recent text explores this methodology in greater depth (Hilborn and Mangel 1997). The resolution of the problem of multiple causality in ecology remains an important task whose achievement may be conditional on the availability of a general approach to the hierarchy of ecological systems.

VI. Conclusions and Prospects

Considering integration across paradigms within a discipline, or indeed considering the role of paradigms that affect all or some large portions of a science, raises the questions of sociological constraints on integration and the problem of objectivity in science. Because one component of the concept of paradigm does involve the belief system of a science, the incorrect conclusion can be drawn that science is subjective. Sociological biases from within science and societal biases from outside can constrain integration across paradigms. The constraints come with scholasticism, personality, approach, and societal pressures and rewards. Such constraints on integration across paradigms can be combated by the practice of science as an open system of understanding constructed by a diverse community of practitioners who are engaged in a transformative interrogation of their subject matter. Transformation, or the molding of understanding and the theories on which it depends, emerges from the dialogue between conceptual constructs and the observable world conducted by a diverse community capable of canceling the individual biases within itself.

Larger paradigms than the four disciplinary perspectives (Things-Stuff, Then-Now) affect the conduct of ecology. These include the contrast between equilibrium and nonequilibrium, the contrast between the unilateral control of biota by the environment and the reciprocal control between biota and abiota, and the emphasis on simple versus multiple causality. Note that one extreme of each of these is more inclusive and therefore can accommodate the older view or approach as a special case when useful and appropriate. The first ecological paradigm shift is resolved by an inclusive nonequilibrium approach. The remaining two paradigm contrasts are still far from resolution.

The ideas we have presented in the first eight chapters have outlined the information necessary to evaluate ecological theory in its broadest sense. We hope that the ideas and examples have provided the framework and concepts necessary to answer such questions as these: What is a theory in ecology? Does a particular proposition or principle represent a whole theory or a part of a theory? What role does a specific conceptual device play in theory or in its test? Is either falsification or prediction the highest necessity for successful science? What is understanding in ecology, and how can it be evaluated? In other words, we hope that the material presented so far allows people to evaluate the philosophical and general methodological claims often made about ecology. We have found that often such claims are programmatically narrow, exclude valid scientific activities, are based on superseded philosophical positions, and persist notwithstanding published philosophical criticism that has exposed such weaknesses. We suspect that the depth and extent of such criticism are simply unknown to most ecologists. Exposure of that conclusion within the discipline has been our motivation to this point. The contemporary philosophy of science supports the strategies actually used by ecologists much better than do the old approaches to philosophy.

9

Ecological Understanding
and the Public

---

*"Profound strictures placed on the language of science are imposed by
the requirement that science is public knowledge, internally consistent,
reproducible, and (in its expression) as unambiguous as possible."*
Raymo 1991:179

I. Overview

The preceding chapters have focused on the structure of understanding within ecology and how
it develops and changes. In this chapter, we look outward, to examine how the science of ecology
interacts with society. Our goals are to determine how public knowledge of ecology might be
improved and how this might, in turn, feed back to enhance the practice of ecological science.
The issues we will examine are the nature of scientific understanding, the nature of scientific
conclusions, the quality of scientific statements in the public sphere, the nature of erroneous
complaints about scientific theory, the role of probability, the lag in public knowledge of ecological understanding, and the rights and responsibilities of scientists, the media, and the public
in the interaction between ecology and society. We will indicate several points at which public
misunderstanding can exist, which can alert both the public and ecologists to opportunities and
strategies for improving public knowledge of ecology.

   To this point, the book has examined the dialogue within science itself, including the role of
the scientific community. However, science also has an ongoing dialogue with the larger society.
Because laypeople may not appreciate the mix of constructive and critical interactions within
science, the interaction between science and society may be compromised. Just as there are ways
to improve our science by attending to the interactions between conceptual constructs and
observable phenomena, so we can improve it by attending to the interactions between science
and society.

   Two kinds of interactions encompass all the issues we wish to highlight. Ecology is perceived
as a problem solver for society. This perception may obscure the fact that a strong foundation
in basic ecology is necessary for its successful application to societal problems. Any science must
have a broad base constructed of fundamental knowledge before it can be applied well. The
second issue is that ecology has implications for how people perceive the workings of the natural
world and their place in it. In other words, the science of ecology has implications about common

187

metaphors and images of the structure and function of the natural world, which in turn affect policy and management decisions.

In this chapter, we take the point of view of science in analyzing the interaction between science and society. We leave it to others to consider the viewpoint of society. We focus our analysis on the contrast between the way that scientists think about the dialogue between science and society and the assumptions that the public might make about science and its application. The analysis should improve the ability of scientists to participate in the larger dialogue between science and society. Sometimes we will emphasize the scientific side of the exchange, whereas other times we will spend more effort on the public side. In particular, attention to the science-public exchange will allow scientists to be better at telling laypeople what they need to know about science. Both the process and the content of science may have to be explained to the public. Ultimately, by considering the interaction between science and the public, we hope that ecologists can learn to translate their science more effectively.

The structure of the chapter is built around the perceptions of certain terms used by both scientists and the public, but in very different ways. As examples of terms used differently by scientists and by the public, we can cite "theory," "belief," and "certainty." We must promote awareness of the contrast in scientific and lay meanings. Understanding the differences and explaining the scientific point of view can promote better use of science.

II. Scientific versus Public Concepts of Theory

Scientists use seemingly ordinary, everyday words to describe the process of science. However, the precise scientific meanings of these seemingly innocent and intuitive terms carry differ from their common usage. When such terms are used in the public sphere to describe science, to communicate scientific conclusions, or to evaluate scientific debates, the nature of scientific understanding and scientific theory may be obscured rather than clarified. What should the public know about the scientific process in order to avoid this confusion? The most important feature of the scientific process for the public to know is that science is an open system of understanding.

A. The Scientific Meaning of Theory

Scientific understanding results from a dialogue between our ideas about how the world is structured or works and the things we can measure or observe in the world. Our ideas of how the world is built or functions are made up of sets of simple, complex, or derived concepts and the accepted facts related to them. All these ideas and facts are tied together in an intellectual framework. The framework illustrates the formal logic, or the links of cause and effect, that tie the concepts and facts together. Understanding is the match between the conceptual side and the observational side of the scientific dialogue. As we showed earlier, the conceptual side develops in many steps and stages in parallel to the accumulation of and working with observations. Therefore, the degree of understanding can be better or worse, and it can change. The view of the world expressed in the conceptual constructs, or theory, is intended to be open to the material world. This means that scientific understanding is intentionally mutable. It is meant to improve. The mutability arises from the constant competition among ideas, with the more compelling or more useful ideas taking hold and working their way into the framework of understanding.

B. Objectivity in Science

Change in the status of scientific understanding is not arbitrary or ad hoc. It is objective for several reasons. The components of understanding are explicit. The ideas, facts, and generalizations are spelled out in theories. The assumptions are stated as parts of theories that can be assessed and rejected or replaced if they or the implications derived from them do not match the observed or measured world. The data that become part of theories or that are used to test theories are repeatable or confirmable by other observers and experimenters. The quality and meaning of observations and theory are subject to evaluation by the entire scientific community.

Objectivity in science is ensured by several tactics. First, individual scientists strive to be disinterested and to accept observations and tests that do not confirm their own expectations. However, even if an individual scientist or working group does not recognize a personal or local error, the second aspect of objectivity can come into play. A diverse scientific community, which has different interests, approaches, and styles of analysis and communication, can expose the errors resulting from the bias of an individual or a subset of the scientific community.

We can use the theory of evolution as an example of the scientific process just outlined. What scientists mean by the theory of evolution is a well-confirmed family of models or a system of complex concepts that explains biological diversity in the natural world and how it changes over time. We will examine these characteristics of evolution next.

C. Evolution as an Example of a Complete, Confirmed Theory

The theory of evolution is a family of models because it is composed of several subfields or smaller theories. Some of the main ones are population genetics, paleontology, biogeography, and natural selection. Population genetics addresses the patterns and mechanisms of hereditary changes within populations of organisms. Paleontology collects data on ancient organisms and explains the long-term patterns of the fossil record. Biogeography deals with the spatial distribution of organisms, which is one of the broad patterns of life on Earth. These models form an integrated network connected by logic and evidence. For example, genetic clocks may indicate a split between chimps and the ancestors of humans at about 7 million years ago and so does the paleontological evidence. The combined paleontological and genetic evidence may suggest human origins in Africa and jointly predict the greatest mitochondrial diversity in Africans. Finally, the theory of evolution contains the theory of natural selection, which provides a general law indicating how evolution can be driven by the external environment interacting with the characteristics of organisms. The law of natural selection was described in detail in Chapter 3. It is a universal conditional statement that (1) if members of a biological entity have heritable variability, and (2) if such variability affects their performance relative to an environment, and (3) if they have the capacity for replication in excess of the capacity of the environment to support them, (4) then progeny of those members that vary in closer conformity with that environment will accumulate in subsequent generations.

The theory is well confirmed because within each of these areas is a broad base of observations and a variety of ways that the generalizations of the theories have been tested and confirmed (Eldridge 1999). The three modes of confirmation discussed by Lloyd (1988) — fit, number of independent lines of evidence, and variety of evidence — are especially important in the longer term and larger scale aspects of the theory, such as paleontology and biogeography. Many of the mechanistic aspects of the theory, such as the components of natural selection, are directly testable by experiment, which is one mode of assessing fit (Brandon 1996).

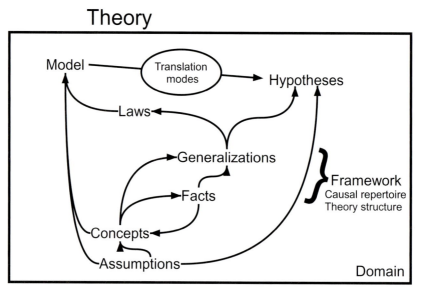

Figure 9.1   Theory is not just a guess. Theory is a complex, interacting system of concepts and empirically based components. Theory contains a large number of components that have functional relationships to one another. This diagram shows one way to conceptualize the general structure of a theory. Note that a theory and its specific components apply to a domain, as indicated by the label inside the box encompassing the components of theory. The framework of a theory is, in essence, the model of how the components of the theory relate to one another. Frameworks also include hierarchical repertoires of the causes that can be drawn upon to generate the specific models and hypotheses. Others schemes for illustrating the relationships between the components of a particular theory are possible. See Chapter 3 for more details on the structure of theory. Note that here we do not separate out definitions but rather leave them as a subset of the application of concepts.

What many laypeople may mean by the theory of evolution is rather different from the scientific meaning just laid out. They often assume the old-fashioned interpretation of the term "theory," which connotes vagueness, tentativeness, and a state of being untested or unproven (Fig. 9.1). Of course, all scientific understanding is, in the final analysis, temporary. New insights, new observations, or new tests can bring a currently confirmed theory into question and invite its replacement, even though a full replacement is rare. However, theories can be confirmed to different degrees, so the best scientific knowledge can consist of confirmed theories — that is, conceptual systems and their established relationships to observed or measurable events and entities. Theory, in the technical sense we have used throughout this book, is the epitome of scientific achievement, and the best theories have a roster of well-developed and explicitly stated components, including laws, confirmed generalizations, confirmable observations and facts, a conceptual framework, models, assumptions, and a clear domain (Chapter 3). The terms that the public often uses to connote certainty (e.g., fact, law) are used in science as components of theory. One cannot do better than a confirmed theory, because such a theory is both inclusive and firm!

One problem with the public view of science is that the dialogue between theory and phenomena often involves debate. Evolutionary theory has been involved in a vigorous debate concerning gradualism and punctuation, for example. Gradualism was a tenet of the classical theory of evolution that assumed trends of change in lineages of species were gradual. In the 1970s, certain paleontologists asserted that the long periods lacking change within the fossil records of many species were not the result of deficiencies in that record but the reality of species change. As a result of sometimes heated discussion, most evolutionary biologists now realize that the long periods lacking change in the fossil record are a legitimate part of the evolutionary pattern (Gould 1989). Debate continues over the exact nature of the changes during the geologically brief (but ecologically long) periods when species do change. The debate within the theory of evolution is the scientific dialogue at work, not the collapse of evolutionary biology or its core! The public perception of evolution would be improved by recognizing the scientific use of common terms, such as theory, and the mutability of scientific understanding. Since changing the public's mind about the meaning of certain terms is unlikely, perhaps the better strategy would be to use modifiers of such publicly misleading nouns as "theory" to assure that the precise scientific meanings were captured in public speech.

III. Certainty and Belief in Science

Presenting scientific conclusions in public discourse has two major components. One is founded on the public expectation of certainty in scientific knowledge, and the second is embodied in contrasting views of belief in the public and scientific spheres.

A. Uncertainty and Change in Science

Certainty can be thought of in two ways. The first view is that certainty is not permanent. Scientific understanding is fundamentally mutable. All scientific understanding is provisional and, therefore, subject to revision. The fact that scientific understanding is mutable means that scientists must honestly say that our current conclusions in any given area are not necessarily the last word. They are only the current last word. When scientists admit this fundamental tenet of science in public, it can lead to frustration and mistrust. Scientists must be better at indicating that although scientific knowledge is provisional, it can be trusted and be well worth acting on (Ziman 2000).

It is important to recognize that change in scientific understanding is only allowed when it is objectively based and founded on fair tests of conceptual constructs against observable phenomena. One does not change a scientific conclusion out of convenience or even an intense desire that things be otherwise. In other words, any certainty in science is necessarily temporary. This is because improvement and refinement are always possible, but change is not arbitrary or subjectively driven.

B. Uncertainty and Probability

Scientific understanding also includes uncertainty that arises from probability and stochasticity. Probability exists in the ecological world because strict determinism is rare. Ecological causes are many, with each cause having only a finite probability of resulting in a particular outcome. This means that the presence or intensity of a certain cause may not always necessarily result in a specific result. Exposing a particular ragweed seed to light and a certain favorable range of temperature in the spring may not stimulate that individual seed to germinate. For whatever reason, genetic or nongenetic, that seed may not respond to what most ragweed seeds would read as a go signal. So the recommendation to treat ragweed seeds in a certain way to cause them to germinate is a probabilistic statement. An ecologist can, on the basis of research with large numbers of ragweed seeds, state that a certain percentage of ragweed seeds will germinate under a specified light and temperature regime. There is an inherent variability in the capacity of ragweed seeds to respond to environmental cues. Incidentally, the probabilistic behavior of ragweed seeds may in fact be an adaptive behavior within this annual, colonizing species. Not all seeds of a cohort will germinate in a given year. Those that do not germinate can typically remain dormant for many years, providing a hedge against the unpredictability of the appearance or quality of disturbed envronments in time and space (Baskin and Baskin 1998).

This example illustrates a widespread situation in ecology. Even when an ecological generalization is derived from a large number of cases, applying the generalization may run afoul of the idiosyncrasies of the specific system or small numbers to which the generalization is applied. One cannot be sure which individual seed will or will not germinate, or which individual mouse will fall prey to an owl on a given night. But one can nevertheless use generalized rates of birth or death in understanding ecosystem structure and change.

There are causes of uncertainty in ecological conclusions other than the inherent variation among individual ecological entities, whether they be organisms, ecosystems, or some other unit. Uncertainty may result from our ignorance of one or more of the factors that might influence some ecological process. For instance, to continue with the ragweed example, if an ecologist worked only with light as a factor in ragweed germination, the capacity to explain the behavior of the seeds would be reduced, because an additional important factor, temperature, is not included in the explanatory repertoire. In addition, to further complicate matters, ecological systems may react to interactions of environmental factors, each of which has its own characteristic uncertainty of impact on the system. Compounding uncertainties is thus a possibility.

C. Belief and Scientific Conclusions

Even when scientists have reached conclusions to reasonable levels of probabilistic understanding and have accounted for multiple important factors, there is a problem in communicating scientific knowledge. The innocent but common statement in the media that "Scientists believe . . ." hides one of the most serious gaps in the way that scientists and the public view the nature of scientific conclusions. In common speech, belief can mean a variety of things, but the use that is furthest from the scientific one is the meaning of a subjective opinion, or an article of faith, which need not be held to any criterion outside the individual.

When applied to science, the term "believe" means something very different. Scientific "belief" means a conclusion that applies to a specific domain and is based on a stated set of assumptions, concepts, definitions, and all the other connected components of theory. Furthermore, if a scientist were to use the term "believe," she would mean that the conclusions were based on having held the theory up against appropriate observations or measurements from the world, which would have tested the match between theory and the world, which could explain new or existing phenomena or could indicate the generality of phenomena observed at certain times and places. Such conclusions would be subject to examination by other scientists who have no stake, or a very different intellectual stake, in the theory of interest. The conclusions can be criticized for the structure of the arguments that have led to them, the veracity of the facts underlying them, and the appropriateness of the tests that have allowed them, so scientific "belief" is a very different creature than personal or societal belief. If, in public discourse, the term "belief" is used to describe a scientific conclusion or the status of a scientific argument, the public must realize that it is not a personal, subjective opinion or faith that is being espoused. Of course, personal belief does have a place in motivating science, but as a pretheoretic notion that is made objective and can be evaluated through the structure and success of theory. The mismatch between the scientific use of "to believe" to mean "to conclude" and public connotation of "to believe" with "to feel," "to guess," or "to take on faith" must be recognized. Incidentally, scientists usually avoid using the term "believe" to prevent the confusion we note here.

D. Acid Rain as an Example

The phenomenon of "acid rain" illustrates some of the problems of differing perceptions of scientific conclusions. The phenomenon is better called "acidic deposition" because dust, snow, and fog, in addition to rainfall, also can deliver acidic chemicals to ecosystems. Often statements such as this are encountered in broadcast or print: "Scientists believe that acid deposition is a serious environmental problem." In fact, there is a strong consensus on the causes, mechanisms, and effects of acid deposition in certain aquatic and terrestrial systems (e.g., Ad Hoc Committee on Acid Rain 1985) so that, on the basis of the scientific foundation, the public connotation of the word "believe" should not be used. Rather, the scientific community can be said to have reached an objective conclusion on the causes, mechanisms, and specific effects of acid deposition. Now, whether an individual or institution chooses to *do* something about reducing the sources of acidification of the atmosphere is a matter of belief, in the common sense. That is, the action or lack thereof has much to do with personal and societal values.

How scientific conclusions are treated in the press reflects not only the public perception of scientific conclusions as a kind of belief but also the media's fundamental desire to be fair. Reporters from various media, following the long-standing journalistic attempt to be balanced, typically try to present opposing views on a subject. For some sorts of scientific matters, this approach can be just the ticket. Many areas of scientific research on the frontier of knowledge do generate various competing alternative theories, hypotheses, models, or other explanations. That is a big part of how science works. The alternatives are sorted, and convergence on a best explanation follows. This process leads scientists to ultimately select the one theory that best fits observations, or has the greatest number of independently tested assumptions, or has the greatest variety of evidence. The process of science, which at a certain point in the history of each subject may legitimately be reported as a raging free-for-all, ultimately can settle down to a correct, or confirmed, theory or to a consensus about a complex issue. When science has settled on some conclusions in a particular area, that should be reported (Likens 1992). Reporters should be aware that the conclusions can change if a theory is found that better fits the data or if new data emerge that threaten an established theory. However, much of scientific understanding that exists at any time will be fairly uncontroversial in the scientific community. In such cases, the journalistic tradition of being fair by seeking out alternative opinions will not accurately reflect the state of the science. This suggests that the reporting of science should follow different models than the adversarial approach taken to many topics.

The hard part for reporters is that it is almost always possible to find a scientist or some other expert to express a dissenting view. One may find people who hold an outmoded theory, who are trying to promote a new alternative theory, or who claim to be in possession of data that demolish an accepted theory. In some cases, these alternative cases will be correct; in others, they will not. It may be that a scientific debate is in an early or ongoing state, or the debate may have been settled to the satisfaction of a diverse or a disinterested community. Reporters should be sensitive to the weight of scientific consensus around a conclusion because science is a system of knowledge and not a supermarket of unconnected ideas competing for attention. If a scientific community has articulated a conclusion about a theory as a result of interrogating that theory carefully and critically, the public should know that. Such consensus is the basis for our successful scientific explanation of the world and for much successful technological and management application of science.

Much of the problem just discussed boils down to the existence of a continuum of acceptance of theories in science (Fig. 9.2; Bauer 1992, Ziman 1978). On one hand is "textbook science," which consists of confirmed theories usually presented in a way that emphasizes their empirical content. Arguments that contradict textbook science are likely to be suspect. On the opposite

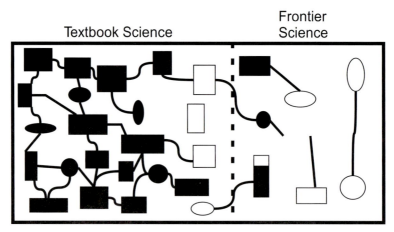

Figure 9.2 Textbook versus frontier science. Textbook science is that body of understanding that is well accepted
and uncontroversial in a discipline. The theories are well developed, well confined, and well integrated.
Frontier science, in contrast, is controversial, provisional, incomplete, and in flux. Firm components of
understanding are indicated by blackened symbols. The different shapes of symbols are intended to suggest
the variety of components of theory. Connections between components indicate a clear framework for
theory. Unconnected lines and open symbols represent tentative or developing aspects of theory.

pole, frontier science is the exciting but relatively primitive attempt to build a new theory or to
conceptually examine and test the components of an emerging area. Here controversy is likely,
because the facts and concepts may be in flux; indeed, even what strands to emphasize in the
dialogue between concept and observation may be arguable. Reporters may save themselves and
their public some confusion by determining whether they are dealing with textbook or frontier
science in a particular instance. Perhaps more subtly, an issue may involve both textbook and
frontier components, so error can be avoided by discriminating between these two aspects of a
case.

How is a layperson to detect valid disagreements versus biased ones about a scientific question?
The first step is to recognize the bases for diverse opinions about a scientific conclusion. The
greatest diversity of opinion may revolve around scientific issues with some societal implications.
That may alert consumers of scientific conclusions to be sensitive to some inappropriate residual
bias among scientists that reflects their social or economic values. Such controversial conclusions
often involve a personal or political decision about acceptable levels of risk or cost of mitigation
of some environmental problem. This, of course, calls on reporters or citizens to apply their own
wariness for possible bias in the news.

An additional way that reporters or the public can evaluate scientific ideas is to be aware that
there are cracked pots in the china shop of science. Often these are experts out of place. The
public has little idea about the great diversity of specialization in contemporary science. A
criticism of the larger theory of evolution by a physiologist might be suspect. It may be that such
criticism focuses on a caricature of the theory rather than on its substance or appropriate application. Criticism of caricatures of theory is a general caution for the public in evaluating how
scientific understanding is used in the public discourse. Criticizing "survival of the fittest" is an
example of an inappropriate caricature, which often draws criticism to the theory of evolution.
The phrase "survival of the fittest" is a metaphor that captures some part of the theory of evolution but it neglects much of the theory and does not even give a sound account of natural

selection. As we stated earlier, natural selection in its technical application is a law in the form of linked conditional statements, each of which can be tested empirically.

An additional check for cracked pots is to look for experts without a theory. If an expert cannot give at least an outline of the theory — in the rigorous sense outlined in this book — from which the expert has derived his or her explanations or forecasts and its connections to other accepted theories, then those explanations or forecasts should be viewed with great suspicion. Unless a supposedly "scientific" forecast of an earthquake on such and such a fault at such and such a time can be tied to a theory and specific data and models, then public action on that forecast is not justified.

The final place to direct a critical public eye is toward experts on the payroll. We certainly do not imply that experts paid for their conclusions in a particular situation are necessarily suspect, but such situations may merit checking for bias. Scientific debate in the public arena may be difficult to evaluate because the content is so far removed from the experience and knowledge of most people. However, there are some features that valid public statements about science should have. This discussion grows out of points made earlier.

IV. Judging Science in the Public Sphere

What kinds of features should scientific information in the public discourse have? How can laypeople know when they are dealing with sound science? There are several general answers, which we outline next.

A. The Presence of a Theory

Theory is the conceptual glue that holds an area of science together. It makes clear the assumptions and limits of discourse and spells out what the phenomena and processes of interest are, how they work, and what their effects are. A theory spells out the implications of the phenomena for the workings of the material world and organizes the accepted facts and generalizations about the phenomena. In the absence of a theory, either a rudimentary or a well-developed one, it is not possible to evaluate empirical claims and their bases.

The safest public discussions of science will be those about areas for which all the components of theory are present. The further from completeness or development an area is, the closer discussion moves to the frontier of a science where controversy and uncertain connections between assumptions, facts, hypotheses, and models exist. Ultimately the public should know the status of a theory, or a family of theories, that underwrites an area and the public discussion of that area.

There are two caveats about using theory in public debate. Sometimes scientists use a brief label of a theory as shorthand for the entire theory. If the theory is controversial, or if the relationships of various parts of a theory are complex, then there is the danger that a shorthand label could be interpreted by different people to represent different aspects of the theory of an area. Scientists should be willing to specify just what subset of the theory of an area they are focusing on. This is especially the case when the subject at hand elicits controversy or excitement in the public sphere. Evolution comes to mind here. It is shorthand for a large, complex, and well-confirmed theory. The shorthand label of "evolution" hardly does the theory justice. The label may indeed be used by some people exploiting the social controversy to cast doubt rather than to suggest the scientific certainty that does exist.

The second caveat about public discussion of theory is the use of metaphors to describe theory. Sometimes the shorthand labels can be made more memorable or cogent if a metaphor is adopted. However, metaphors hold a danger of their own, because they can invite cultural or personal biases into discussion about scientific theory. One of the principal jobs of theory is to permit discussion of a phenomenon to be as exact and free of bias as possible. Phrases like "survival of the fittest" to refer to some or all of the theory of evolution or "the balance of nature" to refer to aspects of population, community, or ecosystem theory introduce a large degree of vagueness into discussions. These metaphors are too broad and contain too much cultural baggage to be useful tags for scientific theories. Because almost all scientific terms also have a metaphorical dimension (Pickett and Cadenasso 2002), this caution is widely applicable.

B. Soundness of Theory

The second criterion for using theory in public discussion is how sound the theory is. Sometimes a theory that is relatively new or untried is the only one available to support the science in an area that the public wants to know about. However, if relatively mature theories are available, then it is possible to evaluate their soundness. Theories that have survived many different tests and have accumulated diverse attempts at confirmation are more likely to be sound bases for public discourse. Furthermore, if discussion is founded on a theory that is clearly related to other confirmed theories, then public confidence in the subject should be increased. For example, dating of fossils or measuring trophic transfers of energy often relies on radioactive isotopes whose properties and behavior are well described by theories of physics and verified by multiple uses in energy production, medicine, or satellite technology. Indeed, conclusions based on such theories inspire confidence among the public because of the wide awareness of their congruence with practice. Note that theories that contradict established well-confined theories are less or not at all trustworthy, although as knowledge develops and the new theory is carefully examined, such apparently contrary theories may eventually be shown to be correct. The public should also be assisted to discriminate between arguments about the periphery or frontier of a science and those about the core of a theory. If a theory is well confirmed, its core is likely to be resistant to attack in the absence of a radically alternative theory that supports better general explanation.

C. The Content of Science in the Public Sphere

There are some standard kinds of arguments about theory that appear in the public sphere. The public should be suspicious of certain kinds of complaints about science in general or about a particular theory. These issues follow.

1. Single Process Definitions of Science

In public discourse, disputants may seek to discredit their opponents by attacking the nature of the science they employ. Often such attempts at dismissal rely on narrow views of what science is supposed to do or what theory is. Recall that we have extracted from the literature a broad view of the goals and structure of science. Such a broad view, based on contemporary philosophy of science, raises suspicions about narrowly based dismissals of science.

Falsification is sometimes taken as the criterion for justifying science. Certainly all science rests on testability, but testability is more than just the potential for falsification. Falsification is a criterion for discriminating science from nonscience that is based on a view of scientific theory as a series of deductively linked statements. The assumption that science is only a logical pursuit, in which the form of arguments is paramount, leaves potential falsification as the only valid approach to science. However, that philosophical assumption has been replaced by a more empirically based view of science (Chapter 1). The empirical nature of science not only allows but actually requires that other modes of evaluating claims made by theory be used. Testing via confirmation involves the attempt to fit theory to the world, the examination of the variety of independently tested aspects of theory, and the evaluation of the number of kinds of evidence brought to bear on a theory. These empirical evaluations go well beyond the simple, logically based approaches to the philosophy of science that formerly dominated the field.

Prediction is also a single criterion sometimes raised as the *sine qua non* of science. This is another holdover from the formerly dominant philosophy of science. Contemporary philosophy holds that the jobs of science are multiple and that no single narrow criterion can characterize science as a whole. Explanation, generalization, and other large goals, along with a variety of more specific tactics such as testing, are also and more inclusively characteristic of science. Ecologists themselves often invoke poor specific predictive power to support their criticisms of models or other new theoretical constructs (Cooper 2003). This criticism does not do justice to the range of goals models serve. Only some models or theory components may legitimately be required to yield specific local predictions, while others may serve many other goals or perform other tasks in generating understanding (see Chapters 4 and 5).

Experimentation is also a process sometimes said to be required for scientific legitimacy. When one can conduct an experiment, it is a desirable thing to do, but there are many situations in which it is unethical or impossible to conduct an experiment. In addition, experiments require simplifying systems to control the various factors that might compromise their outcomes. Experiments are also often conducted over limited temporal and spatial scales because of logistical constraints. So experiments are simplified physical models of the real world that are particularly amenable to query. Equating scientific "proof" with the existence of an experiment is unreasonable, however.

Together, the three cautions we have made about single criterion definitions of what is acceptable as science come down to a single problem: What is proof? The public expects science to provide proofs. Closed deductive systems, such as geometry or math, can be said to provide proofs, but once one moves to an open system of understanding in which the match between reality and concept is the issue, as is the case in the empirical sciences, the requirement of proof as used in geometry or math is inappropriate. Indeed, the classical philosophers, taking their cue from the logic of closed systems, said that the only deductively valid stance toward proof in an empirical science was *dis*proof. We have seen that converting our attention from proof to disproof is not a requirement that must be made of science. If "proof" is to be used in science, it necessarily will be a different beast than proof in closed systems. Scientific "proof" is a multistranded strategy of examining the match between conceptual constructs, or theories, and the way the world is. There is no single method to which science can be reduced. Furthermore, any proof provided by science comes in degrees — a claim can be very well, or well, or just reasonably proven compared to any counterclaim directed at it.

2. Single Cause Explanations

Cause in ecology is often multiple. No single cause is sufficient to explain ecological phenomena. For example, even the common phenomenon of vegetation community change involves, for example, physical features of the site, the availability of organisms, and the interaction of organisms with one another and with the site. Although it may be possible to identify some cluster of specific agents and interactions within this multifactorial universe as affecting community organization, it would be a rare case indeed that was controlled by only one of them. Multiple causes must be invoked for adequate explanation, and ranking causes, or determining which causes act together or antagonistically, is a more reasonable goal than identifying a single dominant cause.

a. Contingency

Ecological causes also involve an important time dimension. What happens in a system, such as a population, community, or landscape, often depends on what has happened there in the past. In cases in which the past matters, ecological systems are said to be contingent. Contingency includes the role of history, the action of trigger factors, the persistent effects of rare events, and the disposition of the system toward certain responses out of all those possible.

b. Scale

It is important for the public to understand that multiplicity of causes is commonly observed because various causal agents occur at different, often nested, spatial and temporal scales. For example, vegetation dieback may be due to chemical leaching due to low rain pH, which may reflect local weather patterns, as well as regional industrialization. Moreover, some soil types in an area may buffer against acid leaching better than others. Each of these factors is a contributory cause, but each reflects a different scale of origin.

3. Universal Statements

Contingency in ecological systems means that there will be few statements (generalizations, hypotheses, laws, predictions, or forecasts) that apply to all ecological systems (Cooper 2003, Lawton 1999). Therefore, the public should be suspicious of ecological pronouncements or arguments that apply to an undivided universe of discourse. Ecological systems have inherent or historically generated features that subdivide the ecological universe. Certainly, no ecological principle will literally apply to the entire universe, because the domain of ecology includes only some of the entities and processes that occur in the universe. While there may be some ecological principles that apply to all systems in the universe *of discourse* that ecology defines, even that smaller conceptual universe may be too large for many statements to cover.

Ecology must employ a taxonomy of cases to divide its universe of discourse into smaller regions or domains in which its principles — laws, generalizations, models — or hypotheses apply. Without a well thought out taxonomy of cases, almost any ecological statement will fail because it overgeneralizes. Expectations should be applied, for example, to heterotrophic species rather than to all species, to nitrogen-limited rather than water-limited systems, or to communities structured by dispersal rather than those structured by local interactions. Such divisions of the universe should be sought to ensure that discussions are productive and that disagreements are legitimate.

The existence of contingency in ecological systems often requires ecologists' answers to include the phrase, "It depends." Such a phrase indicates that a taxonomy of cases needs to be invoked and that the causes of contingency must be recognized. However, ecologists are sometimes too flippant in their use of the phrase. Although the phrase "It depends" can stand as a grammatically complete sentence, it should not be allowed to stand as an intellectually complete sentence. It should, instead, be an introductory clause, after which an ecologist states clearly on what the outcome depends. The factors involved, their interaction, and an estimate (perhaps loose) of the probabilities associated with the statement should be a part of the answer. Explaining contingency to the public and instilling respect for that contingency is an important job that ecologists have not, by and large, done well.

V. The State of Public Knowledge of Ecology

Public knowledge of ecology is out of date in some important ways. Perhaps the most troubling aspect of this situation is that the large metaphors the public uses to carry ecological knowledge into its discourse reinforce the gap between public and scientific understanding of ecology (Kolasa and Pickett 2005). Most people, including some ecologists, view ecology from the perspective of the old paradigm of the discipline (Chapter 7). The old paradigm considered ecological systems to be closed, self-regulating, and existing at or close to point equilibrium. Because of its relative emphasis on equilibrium and its attainment, the classical paradigm has been called the equilibrium paradigm. Its practical implications are that any unit of nature can be understood by studying the dynamics that arise within it, that any such unit is conservable, and that humans are not regular ecological agents. In the public mind, the equilibrium paradigm is closely related to the idea of the "balance of nature," which emphasizes self-regulation, a point of stability, and the exclusion of historical human influences. The balance of nature is not a scientific theory or concept but is a metaphor and cultural palimpsest with deep roots (Egerton 1973). It has apparently supported the persistence of the classical paradigm in ecology (Wu and Loucks 1995).

However, the accumulation of long-term data on the function of many ecological systems brought the hidden assumptions supported by the equilibrium paradigm into view and, hence, into question. The new ecological paradigm accepts that at some temporal and spatial scales, ecological systems may be essentially closed, self-regulating, and equilibrial, but that at other scales (1) ecological populations, communities, ecosystems, landscapes, and so on, can be quite permeable to fluxes of energy, materials, and information from outside them; (2) their regulation can result from frequent disturbances or incursions of competitors or consumers from outside; (3) episodic changes in resource availability occur; or (4) equilibrium, if it exists at all, may be a case of metastability — that is, patch dynamic stability — and require large spatial scales to operate. Such metastability may be based on dynamics and shifts at finer scales. Of course, humans may have a large role to play, given the points of the new paradigm (McDonnell and Pickett 1993).

The metaphor of the balance of nature is a tool for turning these new ecological insights back into the public sphere. There is simply no balance of nature — there is no single reference point of ecological repose enforced by instantaneous, local feedbacks. Nature is highly dynamic, and the status of various ecological systems and their components is determined by accidents of history including human influences, changes in climate, and other environmental conditions, including other organisms and their products. We do not argue that ecological systems do not *tend* to match their environments. They may do so to varying degrees. In fact, determining the mechanisms, rates, and degrees of match is one of ecology's central jobs. The nonequilibrium paradigm merely indicates that the match may be transient, imperfect, and constrained by history and circumstance. Using an ideal equilibrium point to help understand the nature and degree of the tendency is a useful strategy. However, it is not to be taken literally in all cases.

A. The Flux of Nature

The new paradigm suggests a new metaphor, the "flux of nature," that may be useful in alerting the public to the openness, multiplicity of control, lack of equilibrium, and intimate involvement of humans that characterize so much of nature. Of course, there are limits to the degree and rate with which ecological systems can respond to human-caused insults. Nature is neither limitlessly resilient nor perfectly resistant. Just because the natural world is in a state of constant or episodic flux does not mean that any additional human-caused changes are excused. In fact, if those changes exceed (1) the physiological limits of organisms or ecosystem metabolism (i.e., are toxic or stressful), (2) the rates of population growth, (3) the ability of communities or landscape components to reestablish, (4) the rates of dispersal necessary for replacement of organisms or resources to a damaged site, or (5) the limits or genetic variability or recombination to construct newly adapted populations (Pickett et al. 1992), they pose a serious threat to natural resilience and sustainability. Anthropogenic changes of this sort must be allowed only under the most unusual circumstances as they will result in the major reorganization of an ecological system. The flux of nature requires great caution and knowledge about the functioning of the natural world.

In contrast, the "balance of nature" idea has underwritten two opposite views that can both lead to inappropriate management or conservation strategies for ecological systems. First, the "balance of nature" idea can be used to justify any sort of human impact, if nature is considered to be able to return to balance regardless of the insult. Second, the balance of nature can be used to suggest that no human change is justified since it will upset the balance. Neither of these extremes is useful for environmental management or decision making. The "flux of nature" metaphor reminds us that the truth is between these two poles and that careful decisions based on the ecological and evolutionary capacities of systems are required.

The catastrophic fires that occurred in Yellowstone National Park in 1988 illustrate how the public perception of ecology relates to management and policy. The "balance of nature" metaphor suggested that a large area of wilderness such as Yellowstone was self-regulating and in equilibrium. A historical assumption held that North America was essentially thinly occupied and, hence, was little impacted by humans before colonial and industrial society became established in this hemisphere.

Both of these assumptions ended up misleading managers and policy makers (Wagner and Kay 1993). One problem was the management of the elk herd in Yellowstone. Assuming that plants and their herbivores are linked in a "balanced" regulatory feedback and that Native Americans had no role in the landscape that would become part of the park, managers took a passive approach to elk populations. Of course, several managers and researchers noted the increasing damage to the range produced by expanding elk populations and attempted to actively manage the elk population. Other policy needs and perceptions limited elk management, however. More recently, the role of the preindustrial Native American populations in managing range and large ungulate populations has come to the fore (Wagner and Kay 1993). The bottom line is that elk and other large animal populations must be managed, given the absence of the indigenous human use of those herds and the absence of feedbacks between the elk and plant communities that operate to produce a desirable condition of the plant communities.

The second problem to be illustrated by the Yellowstone case is the role of natural disturbance in many ecological systems. Yellowstone was established as a museum of sorts to preserve a sample of the North American landscape from before the machine era. The ecological insight that natural systems are dynamic, even in some cases over huge areas, was not incorporated into the management scheme for Yellowstone. In fact, the history of the area reveals that large fires had a role in structuring the vegetation over the centuries before the park was established. The landscape enshrined in 1884 was the result of the underlying physical environment, of course, but it was also the result of climate periodicity and a related temporal and spatial pattern of fires, in addition to the use of the area by Native Americans, mentioned earlier. Preventing fires led to the removal of one of the main structuring agents of the landscape. It may also have contributed to the severity of the fires once they did start because of the buildup of fuel in excess of the level that would have existed after a series of smaller, periodic fires. Of course, the large fires of 1988 were in part driven by the deep and persistent drought that preceded them, and they would have been difficult if not impossible to prevent. Such large fires are, in any event, an episodic part of the history of certain sorts of forest, such as those in the Yellowstone region.

There is a well-confirmed theory of natural landscapes (Pickett and Rogers 1997, Turner et al. 1993) that embodies the insights needed to manage areas such as Yellowstone. What is lacking is the specific translation mode and data appropriate to many specific sites, but the overall outlines that apply to a large number of ecological systems are available from the theory. The relevant theory has several tenets. Three deal with the basic processes organizing communities. First, systems are structured by site condition, including those generated as a result of disturbance. Second, the availability of propagules (seeds or vegetative means of spread and reproduction) determines the identity of the species that are present in the system. Finally, the interactions among the organisms (including predators, diseases, mutualists, and humans) structure the communities and affect ecosystem functions. The systems are organized as patches or other spatially distinguishable units that are determined by time since last disturbance, the underlying physical environment, and the prevailing climate. Patches change as a result of new disturbances and the other community organizing processes mentioned earlier. The histories of many systems that contain some lasting structural elements (e.g., standing trunks, downed logs, pollen deposits) show the important role of episodic events and disturbances. Note that the term "disturbance" in its ecological use connotes neither negative nor positive value. In ecology, disturbance just *is*, and some organisms and system components are increased as a result, whereas others are decreased (Kolasa 1984). The value accrues to disturbances only through our societal or personal preferences for components, processes, or states of a system that are affected by disturbance.

All this means is that the forests at Yellowstone are mosaics of patches generated in part by the local disturbance by fire. Intense fires have burned in the system as a result of lightning and preindustrial Native American use. Intense fires are unavoidable and result from the intersection of severe drought and fuel status of the forests with a factor of uncertainty resulting from winds generated and changed by the convection and advection of the fires themselves. The regrowth of forest is ensured by the presence of species that are adapted to the open conditions resulting from fires and the reproductive behavior of the forest trees. Knowledge of ecological theory then helps put even catastrophic events in their larger spatial and temporal context and can help inform effective or appropriate response.

VI. Rights and Responsibilities in Ecological Understanding

Ecological understanding, like any other mode of scientific or other kind of understanding, has associated rights and responsibilities. The rights and responsibilities of each constituency that has an interest or stake in ecological understanding differ. Here, we outline those rights and the attendant responsibilities for the public, the media, and ecologists themselves. The rights of one constituency entail responsibilities in the others, so the stakes of the public, the media, and ecologists in the process and products of ecological science are closely intertwined.

A. The Public

People in general have a right to know what it is that ecologists understand and how they come to those conclusions. The public will rightfully wish to apply the knowledge and wisdom of ecology in political discourse and to achieve societal goals.

These two rights suggest several responsibilities that the public has toward ecology. Basically, if ecology is to be most useful and stimulatory to public discourse, it must be unfettered. Ecological understanding must be nurtured in such a way that it can be full and diverse. The public must realize that limiting resources and limiting the questions that ecologists are permitted or supported to pursue will limit the scope and strength of the overall edifice of ecology from which the public wishes to draw information or perspectives. Basic science provides the broad base and the variety of perspectives that are impossible to predict on the basis of societal needs at a particular historical moment but yield the unanticipated knowledge or insight that society requires at some later time. Of course, we realize that resources to support any one science or science as a whole are limited, but a combination of reasonable generosity and unimpeded intellectual freedom yields, in our view, the best science.

One of the most urgent responsibilities of the public is to recognize the difference between scientific understanding, of which ecological understanding is but one variety, and "street" or commonsense modes of knowing. Ecology is in an unfortunate position today because it is synonymous — in the public mind — with political movements or personal philosophies. Environmentalism, "green" political parties, and "deep ecology" all have laid claim to the label for our discipline. There is little to be gained from seeking to reclaim the exclusive scientific pedigree of the term, which dates to Haeckel's (1866) scientific writing. However, the public must realize that the science is not the same as the political or philosophical movements and that individual ecological scientists may or may not hold the political, lifestyle, or spiritual values espoused by parties, movements, ideologies, cells, or congregations in which the word "ecology" is used (Likens 1992). Neither does the scientific understanding and knowledge necessarily underwrite these various political, personal, or social positions. Indeed, some of them are more closely akin to the mythological and prescientific position of the "balance of nature" (Egerton 1973).

The other responsibilities of the public relate to recognizing the difference between the scientific and the common or street conceptions of ecology. Some of these responsibilities have been spelled out in general earlier in this chapter. These are the key points: Laypeople must recognize that scientific understanding is different than personal or community belief systems based on faith, everyday experience alone, or judicial views of truth, for example. Although the term "understanding" can be used for each of these other realms, the structure and origin of those kinds of understanding are completely different from that of science, although they are no better or worse. They simply do different jobs in personal and community life.

Given the difference between science and the other modes of understanding, the public must realize that ecologists mean something specific and reliable by "theory." Theory is the tool for structuring, communicating, and using knowledge in science. It is not, as we have already emphasized, a vague or gauzy matter at all. It is, of course, tentative in the sense that it is subject to alteration on the basis of new objective data, tests, and conceptual refinements. Barring those novel alterations, an established theory is the soundest form of knowing available at a given time. Reliability increases as theories mature and become well tested or confirmed. Similarly, the public must be aware of the difference between personal or community belief and scientific conclusions and between textbook science and frontier science (Bauer 1992).

Scientific understanding in general and, hence, ecological understanding in particular involve probability and multiple causes for events and structures (Kolasa and Pickett 2005, Li 2000). These aspects of science are perhaps the most difficult to grasp, because personal experience and common sense see the world rather differently than does a scientific view. In addition, although individuals judge their own risk in countless situations and decisions every day, personal perceptions of risk and the actual probabilities are often at great variance. The probabilistic viewpoint that characterizes all of contemporary science is perhaps the least well-understood feature of science by nonscientists. At the least, we hope the public can begin to develop some patience for multiple causality and probability.

B. The Media

The media have a right to access scientific information, but that right is inextricably linked to a suite of responsibilities. Reporters must understand the nature of the scientific process. Reporters share many of the misconceptions of science held by the public: for example, it is a body of fact, scientific understanding is fixed and immutable, and causality is singular and deterministic. Reporters are obliged to know even better than the public they serve that science is an open system of understanding, that it is mutable, and that it develops as a result of interrogation of theory and observation by a diverse community of practitioners. The interrogation is itself an open process that enhances objectivity and exposes error. It is especially necessary for reporters to know how scientific conclusions are derived and how consensus about these conclusions is reached. Without such knowledge, reporters risk giving too much weight to contrary opinions, some of which may not be based on confirmed theory, some of which may contradict confirmed theories in related areas, and some of which may represent individual biases, wishful thinking, or crackpot ideas. Contrary ideas and data are critical for the success and progress of science, but not all contrary ideas are appropriate for either the scientific community to accept as confirmed or for the public to act on.

One of the most important obligations of the media is to educate the public about the nature of science. The nature of science is indeed news to almost everybody who is not a scientist. Understanding the basic nature of science is critical to permit the public to comprehend other news about the implications and use of science in public life. Knowing the nature of contemporary science allows the public to comprehend the news about the origin, resurgence, spread, and treatment of diseases; the origin and solutions to environmental problems; the inevitability of certain natural disasters and their role in shaping the physical and biotic world we depend on for bodily and spiritual sustenance; the nature of global environmental change, which is already upon us; and the contribution of science to sound regulations legislation and public decision making.

Two examples will indicate the improvement of public understanding of natural phenomena of societal interest that could be gained if the media were to incorporate a better knowledge of ecological science. The reporting of the 1988 fires in Yellowstone National Park was, for the most part, ignorant of the historic role of fires in the system and of the dynamics of the system. Such errors of fact are relatively easy to correct, but beyond those errors the media applied a model of reporting fires that was inappropriate to the natural world. The urban fire model was used (Smith 1992). There are several assumptions and deductive links in the model that do not necessarily apply to the natural world. The first assumption is that objects the public is interested in are not intended to burn. Buildings and neighborhoods are intended to remain intact, so if they do burn, there are users or owners who have suffered loss of life, use, or investment. This urban fire assumption is illustrated by the use of phrases such as "acres were destroyed" in referring to wild land fires. The model also requires the victims to be identified and, if possible, their reactions reported. Since the burning of buildings is defined as a loss in the urban reporting model, the event is either unintentional or someone has violated the social contract that values buildings and public safety. In either case, the model requires an arsonist or incendiary agent to be identified so that blame can be placed. The next societal step, restitution, may or may not be of concern as news, but it is an important motivation for the assignment of blame that is part of the reporting model for urban fires. This urban fire model was followed in most cases in the reporting of the Yellowstone fires (Smith 1992), but clearly, the assumptions about urban fires are misapplied to natural systems.

Perhaps reporters should seek a different model for reporting at least certain aspects of natural disturbances. In the case of Yellowstone, natural fires and perhaps certain anthropogenic fires were a part of the history of the system and were responsible for major aspects of the structure of the landscape. Although the fires were natural events driven in part by a periodic severe drought, the assignment of blame was inappropriately visited on certain management or policy officials. The search for victims led often to interviewing tourists who reacted to the loss of scenery or experiences of a desired type. Such a search devalued other kinds of experience about the natural world, such as valuing the experience of certain infrequent but powerful and important natural events or valuing the experience of forest and grassland regeneration after fire. Certainly there were important social and personal losses, but they did not constitute the entire story. It seems that journalistic theory should develop another model for structuring reporting of certain natural disturbances, even those that yield some catastrophic human results.

C. Scientists

This community has the society's mandate to discover and interpret the natural world and generate knowledge in the process. To fulfill that mandate, scientists must have the right to go where their fundamental questions lead them, to express contrary scientific conclusions based on theoretical development or the dialogue between theory and observation. We think that science, with all the potential practical societal benefits that can come from it, works only poorly without this freedom. Granting scientists this freedom is worthwhile to society, as exemplified by comparing the quality of science and its spin-offs in totalitarian and free societies.

The right to follow fundamental questions invokes a responsibility for scientists to actually go where their fundamental questions lead them! In other words, scientists have the responsibility to build the edifice of science as soundly and completely as possible. Of course, limitations of time, technology, and other resources will constrain this responsibility, so a secondary responsibility exists to optimize the conduct of science.

Scientists also have a responsibility to look outward to society. Scientists must educate the public, reporters, and policy makers about the nature of the scientific process and how it differs from other modes of understanding more familiar to laypeople (Pickett 2003). In other words, the scientific community must explain the structure and dynamics of scientific understanding and conclusions. Aspects of the paradigm of modern science, including multiple causality and probabilistic causation, must be explained to the public. Scientists have the responsibility to present the most up-to-date view of their science. This requirement refers not only to data but also to paradigms, which may translate into nontechnical metaphors and to theories. However, in explaining the nature of the most up-to-date science, it is proper to differentiate between textbook and frontier science. In other words, what is confirmed and what is controversial must be labeled. Scientists should be willing to state their theories and the specific components of the theories so the strength and components of scientific conclusions can be evaluated openly. This will allow scientists to say where their conclusions come from, whether it be whole theories, specific data, tests, and so forth. Finally, scientists must say as plainly as possible what their conclusions may contribute to public discourse and to use by managers and policy makers.

VII. What It All Means

Science is one of the most important ways people have of understanding the universe and its material components and phenomena. Some have argued that science in general, and ecological science in particular (Eisley *fide* Egerton 1993), has robbed the world of much of its beauty and mystery. Rather, science has revealed new wonders. Science has made the world larger and has exposed new layers of complexity. Ecology has shown us that there are vast webs of interaction, intricate chains of connection, and lasting histories and echoes of past environments. These new, scientifically exposed layers of the world can have a beauty all their own. So there is, it seems to us, great majesty and poetry reflected in the understanding of the natural world provided by contemporary ecology. Ecological explanations of how the natural world is and behaves can be personally rewarding and satisfying. They alert us to worlds beneath the surfaces we see and still other worlds beyond even our most distant vistas. How these various worlds of interaction work and how messages and materials are exchanged among them are still huge frontiers. These worlds are populated by vast numbers of largely invisible organisms whose influence is still largely unknown. Even the large and conspicuous organisms may move so extensively, or accumulate such long histories, that they present us with great mysteries to solve. Still other mysteries lie at the intersections of ecology with other sciences, including those that study humans and their artifacts. That humans have devised a system of understanding that can capture this variety and make mechanistic and material sense of it is a marvelous accomplishment. Perhaps it is inevitable. Perhaps humans would have gone a little crazy without being able to construct testable, revisable, and objective pictures of the world.

Still, science is more than just another incitement to poetry. It has immense practical importance to society, which is a dividend on a necessarily large capital of basic work and non-mission-oriented understanding. Even when technology does not emerge directly from a specific scientific piece of work, technology and science as two perspectives of the world share much of the same approach and, especially in the contemporary world, much of the same basic training. Sometimes the most basic piece of scientific work motivated by nothing more than raw curiosity has surprising practical value. Who would have thought that decades of poking around for pollen in lake sediments would be useful in confronting the contemporary problem of global change, or that equilibrial to nonequilibrial paradigm shifts in ecology would be useful for land managers? Nonetheless, both of these applications are in fact valuable.

Despite its fascination and application, the scientific mode of understanding is also somewhat nonintuitive. Most people have some intuitive understanding of the other modes of understanding: faith, art, and law. All modes of understanding involve constructing a picture of the world or some part of the world of interest and concern and comparing that picture with reality. In some cases, the goal is simply to become comfortable with reality; in other cases, the intent is to feel less small in the face of immense forces and acts. Sometimes the goal is to establish a personal ownership or view of some event or situation; at other times it is to assign guilt or impute innocence. However, the methods and intent of science in establishing understanding are different, on the whole, from those of the other modes of understanding. The fact that science uses familiar terms to describe what it does and what it concludes hides the fact that the process and meaning are entirely different than street, artistic, religious, or legal connotations of the same terms.

We see that the scientific mode of understanding is unique and fills a niche in understanding that complements the others. Superficially, science shares some terms and approaches with the other modes of understanding, so the uninitiated may assume that science does the same things or that scientific arguments and conclusions may be judged in the same way as other arguments and conclusions that appear in the public discourse. This is not true. If the unique role science plays in generating personal wonder and satisfaction, in expanding societal discourse, and in providing support for public decision making is to be maintained, the process of science must be cherished, nurtured, and well and widely taught. Maintaining science and explaining scientific understanding to the public, and enhancing the use of scientific understanding in public discourse, lay responsibility in several places: the scientific and educational communities, the media, the government, and the public itself, which is the source and the patron of the other specialized communities.

LITERATURE CITED

Abbott, I. 1980. Theories dealing with land birds on islands. Advances in Ecological Research **11**:329–371.

Ad Hoc Committee on Acid Rain: Science and Policy. 1985. Is there a scientific consensus on acid rain? Excerpts from six governmental reports. Institute of Ecosystem Studies, Millbrook, NY.

Alberti, M., J. M. Marzluff, E. Shulenberger, G. Bradley, C. Ryan, and C. Zumbrunnen. 2003. Integrating humans into ecology: opportunities and challenges for studying urban ecosystems. BioScience **53**:1169–1179.

Allen, M. F., J. A. MacMahon, and D. C. Anderson. 1984. Reestablishment of *Endogonaceae* on Mount St. Helens: survival of the residuals. Mycologia **76**:1031–1038.

Allen, T. F. H. 1998. The landscape "level" is dead: persuading the family to take it off the respirator. In D. L. Peterson and V. T. Parker, editors. Ecological scale: theory and applications (pp. 35–54). Columbia University Press, New York.

Allen, T. F. H., and T. W. Hoekstra. 1992. Toward a unified ecology. Columbia University Press, New York.

Allen, T. F. H., and T. B. Starr. 1982. Hierarchy: perspectives for ecological complexity. University of Chicago Press, Chicago.

Amsterdamski, S. 1975. Between experience and metaphysics: philosophical problems of the evolution of science. Dordrecht-Holland, Boston.

Andrewartha, H. G., and L. C. Birch. 1954. Distribution and abundance of animals. University of Chicago Press, Chicago.

Anker, P. 2002. The context of ecosystem theory. Ecosystems **5**:611–613.

Armesto, J. J., and S. T. A. Pickett. 1985. Experiments on disturbance in old-field plant communities: impact on species richness and abundance. Ecology **66**:230–240.

Armesto, J. J., S. T. A. Pickett, and M. J. McDonnell. 1991. Spatial heterogeneity during succession: a cyclic model of invasion and exclusion. In J. Kolasa and S. T. A. Pickett, editors. Ecological heterogeneity (pp. 256–269). Springer-Verlag, New York.

Austin, M. P. 1985. Continuum concept, ordination methods and niche theory. Annual Review of Ecology and Systematics **16**:39–61.

Austin, M. P. 1999. A silent clash of paradigms: some inconsistencies in community ecology. Oikos **86**:170–178.

Ayala, F. J. 1974. Introduction. In F. J. Ayala, editor. Studies in the philosophy of biology: reduction and related problems (pp. vii–xvi). University of California Press, Berkeley.

Baguette, M., and V. M. Stevens. 2003. Local populations and metapopulations are both natural and operational categories. Oikos **101**:661–663.

Barbosa, P., V. A. Krischik, and C. G. Jones, editors. 1991. Microbial mediation of plant-herbivore interactions. J. Wiley & Sons, New York.

Bartholomew, G. A. 1982. Scientific innovation and creativity: a zoologist's point of view. American Zoologist **22**:227–235.

Baskin, C. C., and J. M. Baskin. 1998. Ecology of seed dormancy and germination in grasses. In G. P. Cheplick, editor. Population biology of grasses (pp. 30–83). Cambridge University Press, Cambridge.

Bauer, H. H. 1992. Scientific literacy and the myth of the scientific method. University of Illinois Press, Urbana.

Bazzaz, F. A. 1996. Plants in changing environments: linking physiological, population, and community ecology. Cambridge University Press, New York.

Beckner, M. 1974. Reduction, hierarchies and organicism. In F. J. Ayala, editor. Studies in the philosophy of biology: reduction and related problems (pp. 163–177). University of California Press, Berkeley and Los Angeles.

Begon, M., J. L. Harper, and C. R. Townsend. 1996. Ecology: individuals, populations, and communities, 3rd ed. Blackwells, Oxford.

Belnap, J., C. V. Hawkes, and M. K. Firestone. 2003. Boundaries in miniature: two examples from soil. BioScience 53:739–749.

Bernier, R. 1983. Laws in biology. Acta Biotheoretica 32:265–288.

Berry, R. J. 1989. Ecology: where genes and geography meet. Journal of Animal Ecology 58:733–759.

Berryman, A. A. 1987. Equilibrium or non equilibrium: is that the question? Bulletin of the Ecological Society of America 68:500–502.

Berryman, A. A. 2003. On principles, laws and theory in population ecology. Oikos 103:695–701.

Biggs, H. C., and K. H. Rogers. 2003. An adaptive system to link science, monitoring, and management in practice. In J. T. du Toit, K. H. Rogers, and H. C. Biggs, editors. The Kruger experience: ecology and management of savanna heterogeneity (pp. 59–80). Island Press, Washington, DC.

Black, C. C. 1971. Ecological implications of dividing plants into groups with distinct photosynthetic production capacities. Advances in Ecological Research 7:87–114.

Bohm, D. 1996. On dialogue. Routledge, New York.

Bond, E. M., and J. M. Chase. 2002. Biodiversity and ecosystem functioning at local and regional spatial scales. Ecology Letters 5:467–470.

Bormann, F. H., and G. E. Likens. 1979. Catastrophic disturbance and the steady-state in northern hardwood forests. American Scientist 67:660–669.

Botkin, D. B. 1990. Discordant harmonies: a new ecology for the twenty-first century. Oxford University Press, New York.

Botkin, D. B., J. F. Janak, and J. R. Wallis. 1972. Some ecological consequences of a computer model of forest growth. Journal of Ecology 60:849–872.

Botkin, D. B., and M. J. Sobel. 1975. Stability in time-varying ecosystems. American Naturalist 109:625–646.

Boulding, K. E. 1964. The meaning of the twentieth century: the great transition. Harper & Row, New York.

Boyd, R. 1991. Confirmation, semantics, and the interpretation of scientific theories. In R. Boyd, P. Gasper, and J. D. Trout, editors. The philosophy of science (pp. 3–35). MIT Press, Cambridge.

Boyd, R., P. Gasper, and J. D. Trout, editors. 1991. The philosophy of science. The MIT Press, Cambridge.

Boyden, S. 1987. Western civilization in biological perspective: patterns in biohistory. Oxford University Press, Oxford.

Boyden, S. 1993. The human component of ecosystems. In M. J. McDonnell and S. T. A. Pickett, editors. Humans as components of ecosystems: the ecology of subtle human effects and populated areas (pp. 72–77). Springer-Verlag, New York.

Brandon, R. N. 1981. The structural description of evolutionary theory. PSA 2:427–439.

Brandon, R. N. 1984. Adaptation and evolutionary theory. In E. Sober, editor. Conceptual issues in evolutionary biology: an anthology (pp. 58–82). MIT Press, Cambridge.

Brandon, R. N. 1990. Adaptation and environment. Princeton University Press, Princeton, NJ.

Brandon, R. N. 1996. Concepts and methods in evolutionary biology. Cambridge University Press, New York.

Brooks, D. R., and E. O. Wiley. 1988. Evolution as entropy. University of Chicago Press, Chicago.

Brown, J. H. 1971. Mammals on mountain tops: nonequilibrium insular biogeography. American Naturalist 105:467–478.

Brown, J. H. 1981. Two decades of homage to Santa Rosalia: toward a general theory of diversity. Systematic Zoology 21:877–888.

Brown, J. H. 1995. Macroecology. University of Chicago Press, Chicago.

Brown, J. H., and B. A. Maurer. 1989. Macroecology: the division of food and space among species on continents. Science 243:1145–1150.

Brown, W. L., Jr., and E. O. Wilson. 1956. Character displacement. Systematic Zoology 5:49–64.

Brush, S. G. 1974. Should the history of science be rated X? Science 183:1164–1172.

Bryant, J. P., F. S. Chapin, III, and D. R. Klein. 1983. Carbon-nutrient balance of boreal plants in relation to vertebrate herbivory. Oikos 40:357–368.

Bunge, M. 2003. Twenty-five centuries of quantum physics: from Pythagoras to us, and from subjectivism to realism. Science & Education 12:445–466.

Burke, J. 1985. The day the universe changed. Little Brown, Boston.

Cadenasso, M. L., and S. T. A. Pickett. 2000. Linking forest edge structure to edge function: mediation of herbivore damage. Journal of Ecology 88:31–44.

Cadenasso, M. L., and S. T. A. Pickett. 2001. Effects of edge structure on the flux of species into forest interiors. Conservation Biology **15**:91–97.

Cadenasso, M. L., S. T. A. Pickett, K. C. Weathers, S. S. Bell, T. L. Benning, M. M. Carreiro, and T. E. Dawson. 2003a. An interdisciplinary and synthetic approach to ecological boundaries. BioScience **53**:717–722.

Cadenasso, M. L., S. T. A. Pickett, K. C. Weathers, and C. G. Jones. 2003b. A framework for a theory of ecological boundaries. BioScience **53**:750–758.

Cadenasso, M. L., M. M. Traynor, and S. T. A. Pickett. 1997. Functional location of forest edges: gradients of multiple physical factors. Canadian Journal of Forest Research **27**:774–782.

Cadenasso, M. L., K. C. Weathers, and S. T. A. Pickett. 2004. Integrating food web and landscape ecology: subsidies at the regional scale. In G. A. Polis, M. E. Power, and G. Huxel, editors. Food webs at the landscape level (pp 263–267). University of Chicago Press, Chicago.

Cadwallader, M. 1988. Urban geography and social theory. Urban Geography **9**:227–251.

Cale, W. G. 1988. Characterizing populations as entities in ecosystem models: problems and limitations of mass-balance modeling. Ecological Modelling **42**:89–102.

Campbell, D. T. 1974a. "Downward causation" in hierarchically organized biological systems. In F. J. Ayala, editor. Studies in the philosophy of biology: reduction and related problems (pp. 179–186). University of California Press, Berkeley.

Campbell, D. T. 1974b. Unjustified variation and selective retention in scientific discovery. In F. J. Ayala, editor. Studies in the philosophy of biology: reduction and related problems (pp. 139–161). University of California Press, Berkeley.

Carnap, R. 1966. An introduction to the philosophy of science. Basic Books, New York.

Carpenter, S. R. 1996. Microcosm experiments have limited relevance for community and ecosystem ecology. Ecology **77**:677–680.

Carpenter, S. R. 1998. The need for large-scale experiments to assess and predict the response of ecosystems to perturbation. In M. L. Pace and P. M. Groffman, editors. Successes, limitations, and frontiers in ecosystem science (pp. 287–312). Springer, New York.

Carpenter, S. R. 2003. Regime shifts in lake ecosystems. Ecology Institute, Oldendorf/Luhe.

Castle, D. G. A. 2001. A semantic view of ecological theories. Dialectia **55**:51–66.

Castle, D. G. A. 2005. Diversity and stability: theories, models, and data. In K. Cuddington and B. Beisner, editors. Ecological paradigms lost: routes of theory change (pp. 201–212). Elsevier Academic Press, San Diego.

Caswell, H. 1978. Predator — mediated coexistence: a non-equilibrium model. American Naturalist **112**:127–154.

Chave, J. 2004. Neutral theory and community ecology. Ecology Letters **7**:241–253.

Cherrett, J. M., editor. 1989. Ecological concepts: the contribution of ecology to the understanding of the natural world. Blackwell, Oxford.

Chesson, P. L. 1986. Environmental variation and the coexistence of species. In J. Diamond and T. J. Case, editors. Community ecology (pp. 240–256). Harper & Row, New York.

Choi, J. S., A. Mazumder, and R. I. C. Hansell. 1999. Measuring perturbation in a complicated, thermodynamic world. Ecological Modelling **117**:143–158.

Chorley, R. J., and P. Haggett. 1965. Frontiers in geographical teaching. Methuen and Co., London.

Christensen, N. L., A. M. Bartuska, J. H. Brown, S. Carpenter, C. D'Antonio, R. Francis, J. F. Franklin, J. A. MacMahon, R. F. Noss, D. J. Parsons, C. H. Peterson, M. G. Turner, and R. G. Woodmansee. 1996. The report of the Ecological Society of America Committee on the scientific basis for ecosystem management. Ecological Applications **6**:665–691.

Clark, J. S. 1991. Disturbance and tree life history on the shifting mosaic landscape. Journal of Ecology **72**:1102–1118.

Clements, F. E. 1916. Plant succession: an analysis of the development of vegetation. Carnegie Institution of Washington, Washington.

Cody, M. L. 1966. A general theory of clutch size. Evolution **20**:174–184.

Coffin, D. P., and W. K. Lauenroth. 1990. A gap dynamics simulation model of succession in a semiarid grassland. Ecological Modelling **49**:229–266.

Cohen, I. B. 1985. Revolution in science. The Belknap Press of Harvard University Press, Cambridge.

Colemen, J. S., and C. G. Jones. 1991. A phytocentric perspective of phytochemical induction by herbivores. In D. W. Tallamy and M. J. Raupp, editors. Phytochemical induction by herbivores (pp. 3–45). John Wiley & Sons, New York.

Coleman, J. S., C. G. Jones, and V. A. Krischik. 1992. Phytocentric and exploiter perspectives of phytopathology. Advances in Plant Pathology **8**:149–195.

Coley, P. D., J. P. Bryant, and F. S. Chapin, III. 1985. Resource availability and plant antiherbivore defense. Science **230**:895–899.

Colwell, R. K. 1984. What's new? Community ecology discovers biology. In P. W. Price, editor. A new ecology: novel approaches to interactive systems (pp. 287–296). Wiley Interscience, New York.

Colyvan, M., and L. R. Ginzburg. 2003. Laws of nature and laws of ecology. Oikos **101**:649–653.

Connell, J. H. 1978. Diversity in tropical rain forests and coral reefs. Science **199**:1302–1310.

Connell, J. H., I. R. Noble, and R. O. Slatyer. 1987. On the mechanisms producing successional change. Oikos **50**:136–137.

Connell, J. H., and R. O. Slatyer. 1977. Mechanisms of succession in natural communities and their role in community stability and organization. American Naturalist **111**:1119–1144.

Cooper, G. 1998. Generalizations in ecology: a philosophical taxonomy. Biology & Philosophy **13**:555–586.

Cooper, G. J. 2003. The science of the struggle for existence: on the foundations of ecology. Cambridge University Press, New York.

Cooper, W. S. 1913. The climax forest of Isle Royale, Lake Superior, and its development. Botanical Gazette **55**:1–44.

Cooper, W. S. 1926. The fundamentals of vegetation change. Ecology **7**:391–413.

Crombie, A. C. 1953. Augustine to Galileo: the history of science: A.D. 400-1650. Harvard University Press, Cambridge, MA.

Cronon, W. 1983. Changes in the land: Indians, colonists, and the ecology of New England. Hill & Wang, New York.

Cuddington, K. 2001. The balance of nature metaphor and equilibrium in population ecology. Biology & Philosophy **16**:463–479.

Currie, D. J., and V. Paquin. 1987. Large-scale biogeographical patterns of species richness of trees. Nature **329**:326–327.

Cyr, H., and M. L. Pace. 1993. Magnitude and patterns of herbivory in aquatic and terrestrial systems. Nature **361**:148–150.

Dale, V. H., A. E. Lugo, J. A. MacMahon, and S. T. A. Pickett. 1999. Ecosystem management in the context of large, infrequent disturbances. Ecosystems **1**:546–557.

Damuth, J. 1987. Interspecific allometry of population density in mammals and other animals: the independence of body mass and population energy use. Biological Journal of the Linnean Society **31**:193–246.

Danto, A. 1989. Connections to the world: the basic concepts of philosophy. Harper & Row, New York.

Darden, L. 1991. Theory change in science: strategies from Mendelian genetics. Oxford University Press, New York.

Daubenmire, R. F. 1974. Plants and environment: a textbook of autecology. John Wiley & Sons, New York.

Davis, M. B. 1983. Holocene vegetational history of the eastern United States. In E. H. Wright, editor. Late-quaternary environments of the United States: vol 2, the Holocene (pp. 116–181). University of Minnesota Press, Minneapolis.

Davis, M. B. 1986. Climatic instability, time lags, and community disequilibrium. In J. Diamond and T. J. Case, editors. Community ecology (pp. 269–284). Harper & Row, New York.

Dawkins, R., and D. Dennett. 1999. The extended phenotype: the long reach of the gene. Oxford University Press, Oxford.

de Regt, H. W. 2004. Discussion note: making sense of understanding. Philosophy of Science **71**:98–109.

DeAngelis, D. L., and J. C. Waterhouse. 1987. Equilibrium and non equilibrium concepts in ecological models. Ecological Monographs **57**:1–21.

Diamond, J. M. 1974. Colonization of exploded volcanic islands by birds: the supertramp strategy. Science **184**:803–806.

Downing, J. A. 1991. Comparing apples with oranges: methods of interecosystem comparison. In J. Cole, G. M. Lovett, and S. Findlay, editors. Comparative analyses of ecosystems: patterns, mechanisms, and theories (pp. 24–45). Springer-Verlag, New York.

Drake, J. A., G. R. Huxel, and C. L. Hewitt. 1996. Microcosms as models for generating and testing community theory. Ecology **77**:670–677.

Duffy, J. E. 2002. Biodiversity and ecosystem function: the consumer connection. Oikos **99**:201–219.

Edelman, G. M. 1974. The problem of molecular recognition by a selective system. In F. J. Ayala, editor. Studies in the philosophy of biology: reduction and related problems (pp. 45–56). University of California Press, Berkeley.

Egerton, F. N. 1973. Changing concepts of the balance of nature. Quarterly Review of Biology **48**:322–350.

Ehleringer, J. R., and H. A. Mooney. 1978. Leaf hairs: effect on physiological activity and adaptive value to a desert shrub. Oecologia **37**:183–200.

Ehrenfeld, J., and J. P. Schneider. 1991. *Chamaecyperus thyoides* wetlands and suburbanization: effects on hydrology, water quality and plant community composition. Journal in Applied Ecology **28**:467–490.

Ehrlich, P. R. 1989. Discussion: ecology and resource management — is ecological theory any good in practice? In J. Roughgarden, R. M. May, and S. A. Levin, editors. Perspectives in ecological theory (pp. 306–318). Princeton University Press, Princeton, NJ.

Ehrlich, P., and P. H. Raven. 1964. Butterflies and plants: a study in coevolution. Evolution **18**:586–608.

Eldridge, N. 1985. Unfinished synthesis: biological hierarchies and modern evolutionary thought. Oxford University Press, New York.

Eldridge, N. 1999. The pattern of evolution. W. H. Freeman, New York.

Elser, J. J. 2003. Biological stoichiometry: a theoretical framework connecting ecosystem ecology, evolution, and biochemistry for application in astrobiology. International Journal of Astrobiology **2**:185–193.

Facelli, J. M., and S. T. A. Pickett. 1990. Markovian chains and the role of history in succession. Trends in Ecology and Evolution **5**:27–29.

Fagerström, T. 1987. On theory, data and mathematics in ecology. Oikos **50**:258–261.

Feeny, P. 1976. Plant apparency and chemical defense. Recent Advances in Phytochemistry **10**:1–40.

Feyerabend, P. 1975. Against method: outline of an anarchistic theory of knowledge. Humanities Press, Atlantic Highlands, NJ.

Fiedler, P. L., and J. J. Ahouse. 1992. Hierarchies of cause: toward an understanding of rarity in vascular plant species. In P. L. Fiedler and S. K. Jain, editors. Conservation biology: the theory and practice of nature conservation, preservation and management (pp. 21–47). Chapman and Hall, New York.

Fiedler, P. L., and S. Jain, editors. 1992. Conservation biology: the theory and practice of nature conservation, preservation and management. Chapman & Hall, New York.

Fitter, A. H., and R. K. M. Hay. 1987. Environmental physiology of plants, 2nd ed. Academic Press, San Diego, CA.

Ford, E. D. 2000. Scientific method for ecological research. Cambridge University Press, New York.

Forman, R. T. T., A. E. Galli, and C. F. Leck. 1976. Forest size and avian diversity in New Jersey woodlots with some land use implications. Oecologia **26**:1–8.

Forman, R. T. T., and M. Godron. 1986. Landscape ecology. John Wiley & Sons, New York.

Foster, D. R., and J. D. Aber, editors. 2004. Forests in time: the environmental consequences of 1,000 years of change in New England. Yale University Press, New Haven, CT.

Foster, D. R., F. Swanson, J. Aber, I. Burke, N. Brokaw, D. Tilman, and A. Knapp. 2003. The importance of land-use legacies to ecology and conservation. BioScience **53**:77–88.

Fowler, N. L. 1990. Disorderliness in plant communities: comparisons, causes and consequences. In J. B. Grace and D. Tilman, editors. Perspectives on plant competition (pp. 291–306). Academic Press, San Diego, CA.

Fraenkel, G. S. 1959. The raison d'etre of secondary plant substances. Science **129**:1466–1470.

Futuyma, D. J. 1986. Evolutionary biology. Sinauer Associates, Sunderland, MA.

Futuyma, D. J., and G. Moreno. 1988. The evolution of ecological specialization. Annual Review of Ecology and Systematics **19**:207–233.

Gaddis, J. L. 2002. The landscape of history: how historians map the past. Oxford University Press, New York.

Gaines, M. S., N. C. Stenseth, M. L. Johnson, R. A. Ims, and S. Bodrup-Nielsen. 1991. A response to solving the enigma of population cycles with a multifactorial perspective. Journal of Mammalogy **72**:627–631.

Gasper, P. 1991. Causation and explanation. In R. Boyd, P. Gasper, and J. D. Trout, editors. The philosophy of science (pp. 289–297). MIT Press, Cambridge.

Gaston, K. J. 1994. Rarity. Chapman & Hall, London.

Ghiselin, M. T. 1969. The triumph of the Darwinian method. University of Chicago Press, Chicago.

Gilbert, F. S. 1980. The equilibrium theory of island biogeography: fact or fiction? Journal of Biogeography **7**:209–235.

Ginzburg, L. R., and H. R. Akçakaya. 1992. Consequences of ratio-dependent predation for steady-state properties of ecosystems. Ecology **73**:1536–1543.

Gleason, H. A. 1917. The structure and development of the plant association. Bulletin of the Torrey Botanical Club **44**:463–481.

Golley, F. B. 1993. A history of the ecosystem concept in ecology: more than the sum of the parts. Yale University Press, New Haven.

Goodlett, J. C. 1969. Vegetation and the equilibrium concept of landscape. In K. N. M. Greenidge, editor. Essays in plant geography and ecology (pp. 33–44). Nova Scotia Museum, Halifax.

Gould, S. J. 1977. Ever since Darwin: reflections in natural history. Norton, New York.

Gould, S. J. 1984. Darwin's untimely burial. In E. Sober, editor. Conceptual issues in evolutionary biology: an anthology. MIT Press, Cambridge.

Gould, S. J. 1986. Evolution and the triumph of homology, or why history matters. American Scientist **74**:60–69.

Gould, S. J. 1989. Wonderful life: the Burgess shale and the nature of history. Norton, New York.

Gould, S. J. 2002. The structure of evolutionary theory. Harvard University Press, Cambridge, MA.

Grant, P. R., and T. D. Price. 1981. Population variation in continuously varying traits as an ecological genetics problem. American Zoologist **21**:795–811.

Grene, M. 1984. The knower and the known. Studies in the History and Philosophy of Science **23**:305–331.

Grene, M. 1985. Perception, interpretation, and the sciences: toward a new philosophy of science. In D. J. Depew and B. H. Weber, editors. Evolution at a crossroads: the new biology and the new philosophy of science (pp. 1–20). MIT Press, Cambridge.

Grene, M. 1997. Current issues in the philosophy of biology. Perspectives on Science **5**:255–281.

Griffiths, D. 1999. On investigating local-regional species richness relationships. Journal of Animal Ecology **68**:1051–1055.

Grime, J. P. 1979. Plant strategies and vegetation processes. John Wiley & Sons, New York.

Grimm, V. 1998. To be, or to be essentially the same: the "self-identity of ecological units." Trends in Ecology and Evolution **13**:298–299.

Grimm, N. B., J. M. Grove, S. T. A. Pickett, and C. L. Redman. 2000. Integrated approaches to long-term studies of urban ecological systems. BioScience **50**:571–584.

Groffman, P. M., and G. E. Likens, editors. 1994. Integrated regional models: interactions between humans and their environment. Chapman & Hall, New York.

Grubb, P. J. 1992. A positive distrust in simplicity: lessons from plant defense and from competition among plants and among animals. Journal of Ecology **80**:586–610.

Gruden, R. 1990. The grace of great things: creativity and innovation. Ticknor & Fields, New York.

Gunderson, L. H. 2000. Ecological resilience — in theory and application. Annual Review of Ecology and Systematics **31**:425–439.

Gurevitch, J., S. M. Scheiner, and G. A. Fox. 2006. The ecology of plants, 2nd ed. Sinauer Associates Publishers, Sunderland, MA.

Hack, J. T. 1960. Interpretation of erosional topography in humid temperate regions. American Journal of Science **258**:80–97.

Hacking, I. 1983. Representing and intervening: introductory topics in the philosophy of natural science. Cambridge University Press, Cambridge.

Haeckel, E. 1866. Generalle Morphologie der Organismen: allgemeine Grundzuege der organischen Formenwissenschaft, mechanisch begruendet durch die von Charles Darwin reformirte Descendenz-Theorie., 1–2 ed. Reimer, Berlin.

Hagen, J. B. 1989. Research perspectives and the anomolous status of modern ecology. Biology & Philosophy **4**:433–455.

Hagen, J. B. 1992. An entangled bank: the origins of ecosystem ecology. Rutgers University Press, New Brunswick.

Haila, Y. 1986. On the semiotic dimension of ecological theory: the case of island biogeography. Biology & Philosophy **1**:377–387.

Hall, J. 1993. The iceberg and the *Titanic*: human economic behavior in ecological models. In M. J. McDonnell and S. T. A. Pickett, editors. Humans as components of ecosystems: the ecology of subtle human effects and populated areas (pp. 51–60). Springer-Verlag, New York.

Hannon, B. 1986. Ecosystem control theory. Journal of Theoretical Biology **121**:417–437.

Hanski, I., and M. E. Gilpin. 1997. Metapopulation biology: ecology, genetics and evolution. Academic Press, San Diego, CA.

Hanson, N. R. 1961. Is there a logic of scientific discovery? In H. Feigl and G. Maxwell, editors. Current issues in the philosophy of science (pp. 20–42). Holt, Rinehart & Winston, New York.

Hardin, G. 1960. The competitive exclusion principle. Science **131**:1292–1297.

Harper, J. L. 1982. After description. In E. I. Newman, editor. The plant community as a working mechanism (pp. 11–25). Blackwell, London.

Hay, M. E. 1991. Marine-terrestrial contrasts in the ecology of plant chemical defenses against herbivores. Trends in Ecology and Evolution **6**:362–365.

He, F., and P. Legendre. 2002. Species diversity patterns derived from species-area models. Ecology **83**:1185–1198.

Heal, O. W., and J. P. Grime. 1991. Comparative analysis of ecosystems: past lessons and future directions. In J. J. Cole, G.M. Lovett, and S.E.G. Findlay, editors. Comparative analysis of ecosystems: patterns, mechanisms, and theories (pp. 7–23). Springer-Verlag, New York.

Higgins, C. G. 1975. Theory of landscape development: a perspective. In W. N. Melhorn, editor. Theories of landform development. Proc. 6th Annu. Geomorph. Symp. Publ. Geomorph. SUNY/Binghamton, Binghamton, New York.

Hilborn, R., and M. Mangel. 1997. The ecological detective: confronting models with data. Princeton University Press, Princeton.

Hilborn, R., and S. C. Stearns. 1982. On inference in ecology and evolutionary biology: the problem of multiple causes. Acta Biotheoretica **31**:145–164.

Hill, L. 1985. Biology, philosophy, and scientific method. Journal of Biological Education **19**:227–231.

Hils, M. H., and J. L. Vankat. 1982. Species removals from a first-year old-field plant community. Ecology **63**:705–711.

Hoffmann, R. 1988. Nearly circular reasoning. American Scientist **76**:182–185.

Holland, M. M., P. G. Risser, and R. J. Naiman, editors. 1991. Ecotones: the role of landscape boundaries in the management and restoration of changing environments. Chapman & Hall, New York.

Holt, R. D. 1977. Predation, apparent competition and the structure of prey communities. Theoretical Population Biology **12**:197–229.

Holt, R. D. 1987. Prey communities in patchy environments. Oikos **50**:276–290.

Holyoak, M., M. A. Leibold, and R. D. Holt, editors. 2005. Metacommunities: spatial dynamics and ecological communities. University of Chicago Press, Chicago.

Horn, H. S. 1971. The adaptive geometry of trees. Princeton University Press, Princeton.

Horn, H. S., H. H. Shugart, and D. L. Urban. 1989. Simulators as models of forest dynamics. In J. Roughgarden, R. M. May, and S. A. Levin, editors. Perspectives in ecological theory. Princeton University Press, Princeton, NJ.

Hubbell, S. P. 1979. Tree dispersion, abundance, and diversity in a tropical dry forest. Science **203**:1299–1309.

Hubbell, S. P. 2001. The unified neutral theory of biodiversity and biogeography. Princeton University Press, Princeton.

Hull, D. L. 1970. Contemporary systematic philosophies. Annual Review of Ecology and Systematics **1**:19–54.

Hull, D. L. 1974. Philosophy of biological science. Prentice-Hall, Englewood Cliffs, NJ.

Hull, D. L. 1988. Science as a process: an evolutionary account of the social and conceptual development of science. University of Chicago Press, Chicago.

Humphreys, P. 1989. The chances of explanation: causal explanation in the social, medical, and physical sciences. Princeton University Press, Princeton, NJ.

Huston, M. 1979. A general hypothesis of species diversity. American Naturalist **113**:81–101.

Huston, M., D. DeAngelis, and W. Post. 1988. New computer models unify ecological theory. BioScience **38**:682–691.

Huston, M., and T. Smith. 1987. Plant succession: life history and competition. American Naturalist **130**:168–198.

Huszagh, V. A., and J. P. Infante. 1989. The hypothetical way of progress. American Naturalist **338**:109.

Hutchinson, G. E. 1959. Homage to Santa Rosalia, or why are there so many kinds of animals? American Naturalist **93**:145–159.

Ives, A. R. 2005. Community diversity and stability: changing perspectives and changing definitions. In K. Cuddington and B. Beisner, editors. Ecological paradigms lost: routes of theory change. Elsevier Academic Press, San Diego, CA.

Jacob, F. 1982. The possible and the actual. Pantheon Books, New York.

Janovy, J., Jr. 1985. On becoming a biologist. Harper & Row, New York.

Janzen, D. H. 1968. Host-plants as islands in evolutionary and contemporary time. American Naturalist **102**:592–595.

Janzen, D. H. 1973. Host plants as islands: II competition in evolutionary and contemporary time. American Naturalist **107**:786–789.

Jax, K. 1998. Holocoen and ecosystem — on the origin and historical consequences of two concepts. Journal of the History of Biology **31**:113–142.

Jones, C. G. 1991. Interactions among insects, plants, and microorganisms: a net effects perspective on insect performance. In P. Barbosa, V. A. Krischik, and C. G. Jones, editors. Microbial mediation of plant-herbivore interactions (pp. 7–35). John Wiley & Sons, New York.

Jones, C. G., and J. S. Coleman. 1991. Plant stress and insect herbivory: toward an integrated perspective. In H. A. Mooney, W. E. Winner, and E. J. Pell, editors. Response of plants to multiple stresses (pp. 249–280). Academic Press, San Diego, CA.

Jones, C. G., and R. D. Firn. 1991. On the evolution of plant secondary chemical diversity. Philosophical Transactions of the Royal Society of London on Biological Sciences **333**:273–280.

Jones, C. G., R. F. Hopper, J. S. Coleman, and V. A. Krischik. 1993. Control of systemically induced herbivore resistance by plant vascular architecture. Oecologia **93**:452–456.

Jones, C. G., and J. H. Lawton, editors. 1995. Linking species and ecosystems. Chapman & Hall, New York.

Jones, C. G., J. H. Lawton, and M. Shachak. 1994. Organisms as ecosystem engineers. Oikos **69**:373–386.

Jones, C. G., J. H. Lawton, and M. Shachak. 1997. Positive and negative effects of organisms as physical ecosystem engineers. Ecology **78**:1946–1957.

Jones, G. P., and C. Syms. 1998. Disturbance, habitat structure and the ecology of fishes on coral reefs. Australian Journal of Ecology **23**:287–297.

Jordan, W. R., III. 1993. Restoration as a technique for identifying and characterizing human influences on ecosystems. In M. J. McDonnell and S. T. A. Pickett, editors. Humans as components of ecosystems: the ecology of subtle human effects and populated areas (pp. 271–279). Springer-Verlag, New York.

Jørgensen, S. E., and B. D. Fath. 2004. Application of thermodynamic principles in ecology. Ecological Complexity **1**:267–280.

Joseph, G. 1980. The many sciences and the one world. Journal of Philosophy **77**:773–790.

Karban, R., and J. H. Myers. 1989. Induced plant responses to herbivory. Annual Review of Ecology and Systematics **20**:331–348.

Kareiva, P. 1989. Renewing the dialogue between theory and experiments in population ecology. In J. Roughgarden, R. M. May, and S. A. Levin, editors. Perspectives in ecological theory (pp. 68–88). Princeton University Press, Princeton, NJ.

Keddy, P. A. 1976. Lakes as islands: the distributional ecology of two aquatic plants, *Lemna minor L.* and *L. trisulca L.* Ecology **57**:353–358.

Keddy, P. A. 1987. Beyond reductionism and scholasticism in plant community ecology. Vegetatio **69**:209–211.

Keddy, P. A. 1989. Competition. Chapman & Hall, New York.

Keddy, P. A. 1992. Thoughts on a review of a critique for ecology. Bulletin of the Ecological Society of America **73**:234–236.

Keller, D. R., and F. B. Golley. 2000. The philosophy of ecology: from science to synthesis. University of Georgia Press, Athens.

Kiester, A. R. 1980. Natural kinds, natural history and ecology. Synthese **43**:331–342.

Kingsland, S. E. 1985. Modeling nature: episodes in the history of population ecology. University of Chicago Press, Chicago.

Kirchner, J. W. 2003. The Gaia hypothesis: conjectures and refutations. Climatic Change **58**:21–45.

Knapp, A. K., M. D. Smith, S. L. Collins, N. Zambatis, M. Peel, S. Emery, J. Wojdak, M. C. Horner-Devine, H. Biggs, J. Kruger, and S. J. Andelman. 2004. Generality in ecology: testing North American grassland rules in South African savannas. Frontiers in Ecology and Environment **2**:483–491.

Koertge, N., editor. 1998. A house built on sand: exposing postmodernist myths about science. Oxford University Press, Oxford.

Kolasa, J. 1984. Does stress increase ecosystem diversity? Nature **309**:118.

Kolasa, J. 1989. Ecological systems in hierarchical perspective: breaks in community structure and other consequences. Ecology **70**:30–47.

Kolasa, J. 2005. Complexity, system integration, and susceptibility to change: biodiversity connection. Ecological Complexity **2**:431–442.

Kolasa, J., and S. T. A. Pickett. 1989. Ecological systems and the concept of biological organization. Proceedings of the National Academy of Sciences **86**:8837–8841.

Kolasa, J., and S. T. A. Pickett, editors. 1991. Ecological heterogeneity. Springer-Verlag, New York.

Kolasa, J., and S. T. A. Pickett. 2005. Changing academic perspectives of ecology: a view from within. In M. J. Mappin and E. A. Johnson, editors. Environmental education and advocacy (pp. 50–71). Cambridge University Press, New York.

Kolasa, J., and T. N. Romanuk. 2005. Assembly of unequals in the unequal world of a rock pool metacommunity. In M. Holyoak, M. A. Leibold, and R. D. Holt, editors. Metacommunities: spatial dynamics and ecological communities (pp. 212–232). University of Chicago Press, Chicago.

Kolasa, J., and N. Waltho. 1998. A hierarchical view of habitat and its relation to species abundance. In D. Peterson and V. T. Parker, editors. Ecological scale: theory and applications (pp. 55–76). Columbia University Press, New York.

Kozhov, M. M. 1963. Lake Baikal and its life. Junk, The Hague.

Krebs, C. J. 2001. Ecology: the experimental analysis of distribution and abundance, 5th ed. Benjamin Cummings, San Francisco.

Kuhn, T. S. 1962. The structure of scientific revolutions. University of Chicago Press, Chicago.

Kuhn, T. S. 1970. The structure of scientific revolutions, 2nd ed. University of Chicago Press, Chicago.

Kuhn, T. S. 1977. The essential tension: selected studies in scientific tradition and change. University of Chicago Press, Chicago.

Lakatos, I. 1970. Falsification and the methodology of scientific research programmes. In I. Lakatos and A. Musgrave, editors. Criticism and the growth of knowledge (pp. 91–196). Cambridge University Press, New York.

Lange, M. 2005. Ecological laws: what would they be and why would they matter? Oikos **110**:394–403.

Lawton, J. H. 1983. Plant architecture and the diversity of phytophagous insects. Annual Review of Entomology **28**:23–39.

Lawton, J. H. 1999. Are there general laws in ecology? Oikos **84**:177–192.

Lawton, J. H., and C. G. Jones. 1993. Linking species and ecosystem perspectives. Trends in Ecology and Evolution **8**:311–313.

Lawton, J. H., and D. Schroder. 1977. Effects of plant type, size of geographic range and taxonomic isolation on number of insect species associated with British plants. Nature **265**:137–140.

Leary, R. A. 1985. Interaction theory in forest ecology and management. Kluwer Academic, Dordrecht.

Lehman, J. T. 1986. The goal of understanding in limnology. Limnology and Oceanography **31**:1160–1166.

Leibold, M. A., M. Holyoak, N. Mouquet, P. Amarasekare, J. M. Chase, M. F. Hoopes, R. D. Holt, J. B. Shurin, R. Law, D. Tilman, M. Loreau, and A. Gonzalez. 2004. The metacommunity concept: a framework for multi-scale community ecology. Ecology Letters **7**:601–613.

Lekevicius, E. 2003. Ecosystem evolution: major stages and possible mechanisms. Zhurnal Obshchei Biologii **64**:371–388.

Levin, S. A. 1981. The role of theoretical ecology in the description and understanding of populations in heterogeneous environments. American Zoologist **21**:865–875.

Levin, S. A., and R. T. Paine. 1975. The role of disturbance in models of community structure. In S. A. Levin, editor. Ecosystem analysis and prediction. Society for Industrial and Applied Mathematics, Philadelphia.

Levins, R. 1966. The strategy of model building in population biology. American Scientist **54**:421–431.

Levins, R. 1968. Evolution in changing environments: some theoretical explorations. Princeton University Press, Princeton, NJ.

Lewis, R. W. 1982. Theories, structure, teaching and learning. BioScience **32**:734–737.

Lewontin, R. C. 1974. The genetic basis of evolutionary change. Columbia University Press, New York.

Li, B.-L. 2000. Why is the holistic approach becoming so important in landscape ecology? Landscape and Urban Planning **50**:27–41.

Lidicker, W. Z., Jr. 1988. Solving the enigma of microtine "cycles." Journal of Mammalogy **69**:225–235.

Lidicker, W. Z., Jr. 1991. In defense of a multifactor perspective in population ecology. Journal of Mammalogy **72**:631–635.

Likens, G. E., editor. 1989. Long-term studies in ecology: approaches and alternatives. Springer-Verlag, New York.

Likens, G. E. 1991. Human-accelerated environmental change. BioScience **41**:130.

Likens, G. E. 1992. The ecosystem approach: its use and abuse. Ecology Institute, Oldendorf/Luhe, Germany.

Likens, G. E., F. H. Bormann, R. S. Pierce, J. S. Eaton, and N. M. Johnson. 1977. Biogeochemistry of a forested ecosystem. Springer-Verlag, New York.

Lindeman, R. L. 1942. The trophic-dynamic aspect of ecology. Ecology **23**:399–418.

Lloyd, E. A. 1983. The nature of Darwin's support for the theory of natural selection. Philosopy of Science **50**:112–129.

Lloyd, E. A. 1987. Confirmation of ecological and evolutionary models. Biology & Philosophy **2**:277–293.

Lloyd, E. A. 1988. The structure and confirmation of evolutionary theory. Greenwood Press, New York.

Loehle, C. 1983. Evaluation of theories and calculation tools in ecology. Ecological Modelling **19**:239–247.

Loehle, C. 1987a. Errors of construction, evaluation, and inference: a classification of sources of error in ecological models. Ecological Modelling **36**:297–314.

Loehle, C. 1987b. Hypothesis testing in ecology: psychological aspects and the importance of theory maturation. Quarterly Review of Biology **62**:397–410.

Loehle, C. 2004. Challenges of ecological complexity. Ecological Complexity **1**:3–6.

Lomolino, M. V., and G. A. Smith. 2003. Prairie dog towns as islands: applications of island biogeography and landscape ecology for conserving nonvolant terrestrial vertebrates. Global Ecology and Biogeography **12**:275–286.

Longino, H. E. 1990. Science as social knowledge: values and objectivity in scientific inquiry. Princeton University Press, Princeton.

Lotka, A. J. 1922. The stability of the normal age distribution. Proceedings of the National Academy of Sciences **8**:339–345.

Lovelock, J. E. 1979. Gaia: a new look at life on earth. Oxford University Press, New York.

Lovett, G. M., C. G. Jones, M. G. Turner, and K. C. Weathers, editors. 2005. Ecosystem function in heterogeneous landscapes. Springer, New York.

Lowry, W. P. 1967. Weather and life: an introduction to biometeorology. Academic Press, New York.

Lubchenco, J., and B. A. Menge. 1978. Community organization and persistence in a low rocky intertidal zone. Ecological Monographs **59**:67–94.

Luken, J. O. 1990. Directing ecological succession. Chapman & Hall, New York.

MacArthur, R. H. 1972. Geographical ecology: patterns in the distribution of species. Harper & Row, New York.

MacArthur, R. H., and E. O. Wilson. 1967. The theory of island biogeography. Princeton University Press, Princeton, NJ.

Machlis, G. E., J. E. Force, and W. R. Burch, Jr. 1997. The human ecosystem part I: the human ecosystem as an organizing concept in ecosystem management. Society and Natural Resources **10**:347–367.

MacMahon, J. A., D. A. Phillips, J. V. Robinson, and D. J. Schimpf. 1978. Levels of biological organization: an organism-centered approach. BioScience **11**:700–704.

Mahner, M. 1998. Operationalist fallacies in biology. Science & Education **7**:403–421.

Mahner, M., and M. Bunge. 1997. Foundations of biophilosophy. Springer-Verlag, New York.

Mansson, B. A., and J. M. McGlade. 1993. Ecology, thermodynamics and H. T. Odum's conjectures. Oecologia **93**:582–596.

Marks, P. L. 1974. The role of pin cherry (*Prunus pennsylvanica L.*) in the maintenance of stability in northern hardwood ecosystems. Ecological Monographs **44**:73–88.

Marone, L., and R. G. del Solar. In press. Conjectures and confirmations: a role for imagination and inductive inference in the method of ecology.

Martinez, N. 1991. Artifacts or attributes? The effects of resolution on the Little Rock Lake food web. Ecological Monographs **61**:367–392.

Maurer, B. A. 1999. Untangling ecological complexity: the macroscopic perspective. University of Chicago Press, Chicago.

May, R. M. 1981. The role of theory in ecology. American Zoologist **21**:903–910.

May, R. M., and J. Seger. 1986. Ideas in ecology. American Scientist **74**:256–267.

Mayr, E. 1961. Cause and effect in biology. Science **134**:1501–1506.

Mayr, E. 1982. The growth of biological thought: diversity, evolution, and inheritance. The Belknap Press of Harvard University Press, Cambridge.

Mayr, E. 1988. Toward a new philosophy of biology: observations of an evolutionist. The Belknap Press of Harvard University Press, Cambridge.

Mayr, E. 1991. One long argument: Charles Darwin and the genesis of modern evolutionary thought. Harvard University Press, Cambridge.

Mayr, E. 1996. The autonomy of biology: the position of biology among sciences. Quarterly Review of Biology **71**:97–106.

McCann, K. S. 2000. The diversity-stability debate. Nature **405**:228–233.

McCann, K. S. 2005. Perspectives on diversity, structure, and stability. In K. Cuddington and B. Beisner, editors. Ecological paradigms lost: routes of theory change (pp. 183–200). Elsevier Academic Press, San Diego, CA.

McDonnell, M. J., and S. T. A. Pickett. 1988. Connectivity and the theory of landscape ecology. Munstersche Geographische Arbeiten **29**:17–21.

McDonnell, M. J., and S. T. A. Pickett, editors. 1993. Humans as components of ecosystems: the ecology of subtle human effects and populated areas. Springer-Verlag, New York.

McGill, B. 2003. Strong and weak tests of macroecological theory. Oikos **102**:679–685.

McGrath, B., M. L. Cadenasso, J. M. Grove, V. Marshall, S. T. A. Pickett, J. Towers, and eds. 2007. Designing urban patch dynamics. Columbia University Gradnate School of Architecture, Planning & Preservation, New York.

McIntosh, R. P. 1985. The background of ecology: concept and theory. Cambridge University Press, Cambridge.

McIntosh, R. P. 1987. Pluralism in ecology. Annual Review of Ecology and Systematics **18**:321–341.

McNaughton, S. J., M. Oesterheld, D. A. Frank, and K. J. Williams. 1989. Ecosystem-level patterns of primary productivity and herbivory in terrestrial habitats. Nature **341**:142–144.

Meiners, S. J., and S. T. A. Pickett. 1999. Changes in community and population responses across a forest-field gradient. Ecography **22**:261–267.

Melillo, J. M., J. D. Aber, and J. F. Muratore. 1982. Nitrogen and lignin control of hardwood leaf litter decomposition dynamics. Ecology **63**:621–626.

Merriam, G. 1984. Connectivity: a fundamental characteristic of landscape pattern. In J. Brandt, editor. Methodology in landscape ecological research (pp. 5–15). Roskilde University Press, Roskilde, Denmark.

Mikkelson, G. M. 2001. Untangling ecology? Biology & Philosophy **16**:273–279.

Mikkelson, G. M. 2003. Ecological kinds and ecological laws. Philosophy of Science **70**:1390–1400.

Miles, J. 1979. Vegetation dynamics. Wiley, New York.

Miller, R. W. 1987. Fact and method: explanation, confirmation and reality in the natural and the social sciences. Princeton University Press, Princeton.

Mitchell, S. D. 2002. Integrated pluralism. Biology & Philosophy **17**:55–70.

Morin, P. J. 1984. Odonate guild composition: experiments with colonization history and fish predation. Ecology **65**:1866–1873.

Müller, F. 1997. State-of-the-art in ecosystem theory. Ecological Modelling **100**:135–161.

Murray, B. G., Jr. 1986. The structure of theory, and the role of competition in community dynamics. Oikos **46**:145–158.

Murray, B. G., Jr. 2000. Universal laws and predictive theory in ecology and evolution. Oikos **89**:403–408.

Murray, B. G., Jr. 2001. Are ecological and evolutionary theories scientific? Biological Reviews **76**:255–289.

Naeem, S. 2002. Biodiversity: biodiversity equals instability? Nature **416**:84–86.

Nagel, E. 1961. The structure of science: problems in the logic of scientific explanation. Harcourt, Brace and World, New York.

Nee, S. 1990. Community construction. Trends in Ecology and Evolution **5**:337–340.

Nisbet, R. M., E. B. Muller, K. Lika, and S. A. L. M. Kooijman. 2000. From molecules to ecosystems through dynamic energy budget models. Journal of Animal Ecology **69**:913–926.

Nixon, S. 2001. Some reluctant ruminations on scales (and claws and teeth) in marine mesocosms. In R. H. Gardner, W. M. Kemp, V. S. Kennedy, and J. E. Petersen, editors. Scaling relations in experimental ecology (pp. 179–190). Columbia University Press, New York.

Nonacs, P. 1993. Is satisficing an alternative to optimal foraging theory. Oikos **67**:371–375.

Norris, S. 2003. Neutral theory: a new, unified model for ecology. BioScience **53**:124–129.

Novak, J. D., and D. B. Gowin. 1984. Learning how to learn. Cambridge University Press, New York.

O'Hear, A. 1990. An introduction to the philosophy of science. Oxford University Press, New York.

O'Neill, R. V. 2001. Is it time to bury the ecosystem concept? (with full military honors, of course!). Ecology **82**:3275–3284.

O'Neill, R. V., D. L. DeAngelis, J. B. Waide, and T. F. H. Allen. 1986. A hierarchical concept of ecosystems. Princeton University Press, Princeton.

O'Neill, R. V., and A. W. King. 1998. Homage to St. Michael; or, why are there so many books on scale? In D. L. Peterson and V. T. Parker, editors. Ecological scale: theory and application (pp. 3–15). Columbia University Press, New York.

O'Neill, R. V., K. H. Riitters, J. D. Wickham, and K. B. Jones. 1999. Landscape pattern metrics and regional assessment. Ecosystem Health **5**:225–233.

Odenbaugh, J. 2001. Ecological stability, model building, and environmental policy: a reply to some of the pessimism. Philosophy of Science **68**:S493–S505.

Odling-Smee, F. J., K. N. Laland, and M. W. Feldman. 2003. Niche construction: the neglected process in evolution. Princeton University Press, Princeton.

Odum, E. P. 1953. Fundamentals of Ecology. Saunders, Philadelphia.

Odum, E. P. 1969. The strategy of ecosystem development. Science **164**:262–270.

Odum, E. P. 1971. Fundamentals of ecology, 3rd ed. Saunders, Philadelphia.

Odum, E. P. 1990. Field experimental tests of ecosystem-level hypotheses. Trends in Ecology and Evolution **5**:204–205.

Odum, E. P. 1996. Ecology: a bridge between science and society. Sinauer, Sunderland, MA.

Odum, H. T. 1983. Systems ecology: an introduction. Wiley, New York.

Oliver, C. D., and B. C. Larson. 1990. Forest stand dynamics. McGraw-Hill, New York.

Oosting, H. J. 1942. An ecological analysis of the plant communities of Piedmont, North Carolina. American Midland Naturalist **28**:1–126.

Opdam, P. D., D. van Dorp, and C. J. F. Ter Braak. 1984. The effect of isolation on the number of woodland birds in small woods in the Netherlands. Journal of Biogeography **11**:473–478.

Oreskes, N., K. Shrader-Frechette, and K. Belitz. 1994. Verification, validation, and confirmation of numerical models in the Earth sciences. Science **263**:641–646.

Orians, G. H. 2005. Cumulative threats to the environment. Environment **37**:7–14.

Oster, G. 1981. Predicting populations. American Zoologist **21**:831–844.

Pace, M. L. 2001. Getting it right and wrong: extrapolations across experimental scales. In R. H. Gardner, W. M. Kemp, V. S. Kennedy, and J. E. Petersen, editors. Scaling relations in experimental ecology (pp. 157–177). Columbia University Press, New York.

Pace M. L., and P. M. Groffman, editors. 1998. Successes, limitations, and frontiers in ecosystem science. Springer-Verlag, New York.

Padoch, C. 1993. Part II: a human ecologist's perspective. In M. J. McDonnell and S. T. A. Pickett, editors. Humans as components of ecosystems: the ecology of subtle human effects and populated areas (pp. 303–305). Springer-Verlag, New York.

Paine, R. T., and S. A. Levin. 1981. Intertidal landscapes: disturbance and the dynamics of pattern. Ecological Monographs **51**:145–178.

Palmer, M., E. Bernhardt, E. Chornesky, S. Collins, A. Dobson, C. Duke, B. Gold, R. Jacobson, S. Kingsland, R. Kranz, M. Mappin, M. L. Martinez, F. Micheli, J. Morse, M. Pace, M. Pascual, S. Palumbi, O. J. Reichman, A. Simons, A. Townsend, and M. Turner. 2004. Ecology for a crowded planet. Science **304**:1251–1252.

Palmer, M. W., and P. S. White. 1994. On the existence of ecological communities. Journal of Vegetation Science **5**:279–282.

Palter, R. 1984. Relativity and other issues. Science **226**:684–685.

Parker, V. T. 2004. Community of the individual: implications for the community concept. Oikos **104**:27–34.

Passioura, J. B. 1979. Accountability, philosophy and plant physiology. Search **10**:347–350.

Pattee, H. H. 1973. The physical basis and origin of hierarchical control. In H. H. Pattee, editor. Hierarchy theory: the challenge of complexity (pp. 71–108). Braziller, New York.

Peet, R. K., D. Glenn-Lewin, and J. W. Wolf. 1983. Prediction of man's impact on vegetation. In W. Holzner, editor. Man's impact on vegetation (pp. 41–53). Junk, The Hague.

Peierls, B. L., N. F. Caraco, and J. J. Cole. 1991. Human influence on river nitrogen. Nature **350**:386–387.

Perry, G., and E. R. Pianka. 1997. Animal foraging: past, present and future. Trends in Ecology and Evolution **12**:360–364.

Peters, R. H. 1980. Useful concepts for predictive ecology. Synthese **43**:257–269.

Peters, R. H. 1986. The role of prediction in limnology. Limnology and Oceanography **31**:1143–1159.

Peters R. H. 1991. A critique for ecology. Cambridge University Press, Cambridge.

Peterson, C. H. 1991. Intertidal zonation of marine invertebrates in sand and mud. American Scientist **79**:236–249.

Petraitis, P. S., R. E. Latham, and R. A. Niesenbaum. 1989. The maintenance of species diversity by disturbance. Quarterly Review of Biology **64**:393–418.

Pickett, S. T. A. 1976. Succession: an evolutionary interpretation. American Naturalist **110**:107–119.

Pickett, S. T. A. 1991. Long-term studies: past experience and recommendations for the future. In P. G. Risser, editor. Long-term ecological research (pp. 71–88). Wiley, Chichester.

Pickett, S. T. A. 1998. Natural processes. In M. J. Mac, editor. Status and trends of the nation's biological resources (pp. 11–19). U.S. Department of Interior, U.S. Geological Survey, Reston, VA.

Pickett, S. T. A. 1999. The culture of synthesis: habits of mind in novel ecological integration. Oikos **87**:479–487.

Pickett, S. T. A. 2003. Why is developing a broad understanding of urban ecosystems important to science and scientists? In A. R. Berkowitz, C. H. Nilon, and K. S. Hollweg, editors. Understanding urban ecosystems: a new frontier for science and education (pp. 58–72). Springer-Verlag, New York.

Pickett, S. T. A., W. R. Burch, Jr., and J. M. Grove. 1999. Interdisciplinary research: maintaining the constructive impulse in a culture of criticism. Ecosystems **2**:302–307.

Pickett, S. T. A., and M. L. Cadenasso. 2002. Ecosystem as a multidimensional concept: meaning, model and metaphor. Ecosystems **5**:1–10.

Pickett, S. T. A., and M. L. Cadenasso. 2005. Vegetation succession. In E. van der Maarel, editor. Vegetation ecology (pp. 172–198). Blackwell, Malden, MA.

Pickett, S. T. A., M. L. Cadenasso, J. M. Grove, C. H. Nilon, R. V. Pouyat, W. C. Zipperer, and R. Costanza. 2001. Urban ecological systems: linking terrestrial ecological, physical, and socioeconomic components of metropolitan areas. Annual Review of Ecology and Systematics **32**:127–157.

Pickett, S. T. A., M. L. Cadenasso, and C. G. Jones. 2000. Generation of heterogeneity by organisms: creation, maintenance, and transformation. In M. Hutchings, editor. Ecological consequences of habitat heterogeneity (pp. 33–52). Blackwell, New York.

Pickett, S. T. A., S. L. Collins, and J. J. Armesto. 1987a. A hierarchical consideration of causes and mechanisms of succession. Vegetatio **69**:109–114.

Pickett, S. T. A., S. L. Collins, and J. J. Armesto. 1987b. Models, mechanisms and pathways of succession. Botanical Review **53**:335–371.

Pickett, S. T. A., and J. Kolasa. 1989. Structure of theory in vegetation science. Vegetatio **83**:7–15.

Pickett, S. T. A., J. Kolasa, J. J. Armesto, and S. L. Collins. 1989. The ecological concept of disturbance and its expression at various hierarchical levels. Oikos **54**:129–136.

Pickett, S. T. A., and M. J. McDonnell. 1989. Changing perspectives in community dynamics: a theory of successional forces. Trends in Ecology and Evolution **4**:241–245.

Pickett, S. T. A., V. T. Parker, and P. L. Fiedler. 1992. The new paradigm in ecology: implications for conservation biology above the species level. In P. L. Fiedler and S. K. Jain, editors. Conservation biology: the theory and practice of nature conservation, preservation, and management (pp. 65–88). Chapman & Hall, New York.

Pickett, S. T. A., and K. H. Rogers. 1997. Patch dynamics: the transformation of landscape structure and function. In J. A. Bissonette, editor. Wildlife and landscape ecology: effects of pattern and scale (pp. 101–127). Springer-Verlag, New York.

Pickett, S. T. A., and P. S. White, editors. 1985. The ecology of natural disturbance and patch dynamics. Academic Press, Orlando, FL.

Pielou, E. C. 1981. The usefulness of ecological models: a stock-taking. Quarterly Review of Biology **56**:17–31.

Pimm, S. L. 1984. The complexity and stability of ecosystems. Nature **307**:321–326.

Pimm, S. L. 1991. The balance of nature? Ecological issues in the conservation of species and communities. University of Chicago Press, Chicago.

Platt, J. R. 1964. Strong inference. Science **146**:347–353.

Pool, R. 1989. Is it chaos, or is it just noise? Science **243**:25–28.

Popper, K. R. 1959. The logic of scientific discovery. Hutchison, London.

Popper, K. R. 1965. Conjectures and refutations: the growth of scientific knowledge. Harper & Row, New York.

Popper K. R. 1968. The logic of scientific discovery, 3rd ed. Harper & Row, New York.

Popper, K. R. 1974. Scientific reduction and the essential incompleteness of all science. In F. J. Ayala, editor. Studies in the philosophy of biology: reduction and related problems (pp. 259–284). University of California Press, Berkeley and Los Angeles.

Price, D. J. 1961. Science since Babylon. Yale University Press, New Haven, CT.

Price, P. W. 1984. Alternative paradigms in community ecology. In P. W. Price, editor. A new ecology: novel approaches to interactive systems (pp. 351–383). Wiley Interscience, New York.

Putnam, H. 1975. Mind, language and reality. Cambridge University Press, New York.

Quenette, P. Y., and J. F. Gerard. 1993. Why biologists do not think like Newtonian physicists. Oikos **68**:361–363.

Quine, W. V., and J. S. Ullian. 1978. The web of belief, 2nd ed. Random House, New York.

Quinn, J. F., and S. P. Harrison. 1988. Effects of habitat fragmentation and isolation on species richness: evidence from biogeographic patterns. Oecologia **75**:132–140.

Rappoport, A. 1978. Various meaning of "theory." General Systems **23**:29–37.

Ratner, V. A. 1990. Towards a unified theory of molecular evolution (TIME). Theoretical Population Biology **38**:233–261.

Raymo, C. 1991. The virgin and the mousetrap: essays in search of the soul of science. Viking Press, New York.

Reed, E. S. 1981. The lawfulness of natural selection. American Naturalist **118**:61–71.

Reice, S. R. 1994. Nonequilibrium determinants of biological community structure. American Scientist **82**:424–435.

Reiners, W. A. 1986. Complementary models for ecosystems. American Naturalist **127**:59–73.

Rensch, B. 1974. Polynomistic determination of biological process. In F. J. Ayala, editor. Studies in the philosophy of biology: reduction and related problems (pp. 241–258). University of California Press, Berkeley and Los Angeles.

Ribas, C. R., J. H. Schoereder, M. Pic, and S. M. Soares. 2003. Tree heterogeneity, resource availability, and larger scale processes regulating arboreal ant species richness. Austral Ecology **28**:305–314.

Rigler, F. H. 1982. Recognition of the possible: an advantage of empiricism in ecology. Canadian Journal of Fisheries and Aquatic Sciences **39**:1323–1331.

Risser, P. G. 1987. Landscape ecology: state of the art. In M. G. Turner, editor. Landscape heterogeneity and disturbance (pp. 3–14). Springer-Verlag, New York.

Risser, P. G. 1988. Abiotic controls on primary productivity and nutrient cycles in North American grasslands. In L. R. Pomeroy and J. J. Alberts, editors. Concepts of ecosystem ecology: a comparative view (pp. 115–129). Springer-Verlag, New York.

Ritchie, M. E., and H. Olff. 1999. Spatial scaling laws yield a synthetic theory of biodiversity. Nature **400**:557–560.

Roberts, D. W. 1987. A dynamical systems perspective on vegetation theory. Vegetatio **69**:27–33.

Robertson, A. 1991. Plant-animal interactions and the structure and function of mangrove forest ecosystems. Australian Journal of Ecology **16**:433–443.

Rogers, K. H. 1997. Operationalizing ecology under a new paradigm: an African perspective. In S. T. A. Pickett, R. S. Ostfeld, M. Shachak, and G. E. Likens, editors. The ecological basis of conservation: heterogeneity, ecosystems, and biodiversity (pp. 60–70). Chapman & Hall, New York.

Rohrlich, F. 1987. From paradox to reality: our basic concepts of the physical world. Cambridge University Press, Cambridge.

Romme, W. H. 1982. Fire and landscape diversity in subalpine forests of Yellowstone National Park. Ecological Monographs **52**:199–221.

Rosenberg, A. 1985. The structure of biological science. Cambridge University Press, Cambridge.

Rosenzweig, M. L. 1974. And replenish the Earth: the evolution, consequences, and prevention of overpopulation. Harper & Row, New York.

Rosenzweig, M. L. 1995. Species diversity in space and time. Cambridge University Press, New York.

Rosenzweig, M. L., and Y. Ziv. 1999. The echo pattern of species diversity: pattern and processes. Ecography **22**:614–628.

Roughgarden, J. 1984. Competition and theory in community ecology. In G. Salt, editor. Ecology and evolutionary biology (pp. 3–21). University of Chicago Press, Chicago.

Roughgarden, J. 1989. The structure and assembly of communities. In J. Roughgarden, R. M. May, and S. A. Levin, editors. Perspectives in ecological theory (pp. 203–226). Princeton University Press, Princeton, NJ.

Roughgarden, J., and J. Diamond. 1986. Overview: the role of species interactions in community ecology. In J. Diamond and T. J. Case, editors. Community ecology (pp. 332–343). Harper & Row, New York.

Roughgarden J., R. M. May, and S. A. Levin, editors, 1989. Perspectives in ecological theory. Princeton University Press, Princeton, NJ.

Ruse, M. 1979. Falsifiability, consilience, and synthesis. Systematic Zoology **28**:530–536.

Ruse, M. 1988. Philosophy of biology today. State University of New York Press, Albany.

Russell, E. W. B. 1993. Discovery of the subtle. In M. J. McDonnell and S. T. A. Pickett, editors. Humans as components of ecosystems: the ecology of subtle human effects and populated areas (pp. 81–90). Springer-Verlag, New York.

Russell, E. W. B. 1997. People and the land through time. Yale University Press, New Haven.

Rykiel, E. J. 1985. Towards a definition of ecological disturbance. Australian Journal of Ecology **10**:361–365.

Sagoff, M. 1997. Muddle or muddle through? Takings jurisprudence meets the endangered species act. William and Mary Law Review **38**:825–993.

Sagoff, M. 2003. The plaza and the pendulum: two concepts of ecological science. Biology & Philosophy **18**:529–552.

Salmon, W. C. 1984. Scientific explanation and the causal structure of the world. Princeton University Press, Princeton.

Salthe, S. N. 1985. Evolving hierarchical systems: their structure and representation. Columbia University Press, New York.

Scheiner, S. M. 1994. Why ecologists should care about philosophy: a reply to Keddy's reply. Bulletin of the Ecological Society of America 75:50–52.

Scheiner, S. M., A. J. Hudson, and M. A. Vandermeulen. 1993. An epistemology for ecology. Bulletin of the Ecological Society of America 74:17–21.

Schiebinger, L. 1999. Has feminism changed science? Harvard University Press, Cambridge.

Schimper, A. F. W. 1903. Plant geography upon a physiological basis. Clarendon Press, Oxford.

Schindler, D. W. 1998. Replication versus realism: the need for ecosystem-scale experiments. Ecosystems 1:323–334.

Schlesinger, W. H. 1991. Biogeochemistry: an analysis of global change. Academic Press, San Diego, CA.

Schoener, T. W. 1986a. Mechanistic approaches to community ecology: a new reductionism? American Zoologist 26:81–106.

Schoener, T. W. 1986b. Overview: kinds of ecological communities — ecology becomes pluralistic. In J. Diamond and T. J. Case, editors. Community ecology (pp. 467–479). Harper & Row, New York.

Scriven, M. 1959. Explanation and prediction in evolutionary theory. Science 130:477–482.

Shachak, M., and S. T. A. Pickett. 1997. Linking ecological understanding and application: patchiness in dryland systems. In S. T. A. Pickett, R. S. Ostfeld, M. Shachak, and G. E. Likens, editors. The ecological basis of conservation: heterogeneity, ecosystems, and biodiversity (pp. 108–119). Chapman & Hall, New York.

Shafer, C. L. 1990. Nature reserves: island theory and conservation practice. Smithsonian Institution Press, Washington, DC.

Shapere, D. 1974. On the relations between compositional and evolutionary theories. In F. J. Ayala, editor. Studies in the philosophy of biology: reduction and related problems (pp. 187–204). University of California Press, Berkeley and Los Angeles.

Shipley, W., and P. A. Keddy. 1987. The individualistic and community-unit concepts and falsifiable hypotheses. Vegetatio 69:47–55.

Shrader-Frechette, K. 2001. A companion to environmental philosophy. Blackwell, Malden, MA.

Shrader-Frechette, K. S., and E. D. McCoy. 1993. Method in ecology: strategies for conservation. Cambridge University Press, New York.

Shrader-Frechette, K. S., and E. D. McCoy. 1994. Applied ecology and the logic of case studies. Philosophy of Science 61:228–249.

Shugart, H. H. 1984. A theory of forest dynamics: the ecological implications of forest succession models. Springer-Verlag, New York.

Shugart, H. H. 1989. The role of ecological models in long-term ecological studies. In G. E. Likens, editor. Long-term studies in ecology, approaches and alternatives (pp. 90–109). Springer-Verlag, New York.

Shugart, H. H., and S. W. Seagle. 1985. Modeling forest landscapes and the role of disturbance in ecosystems and communities. In S. T. A. Pickett and P. S. White, editors. The ecology of natural disturbance and patch dynamics (pp. 353–368). Academic Press, Orlando.

Silliman, B. R., and M. D. Bertness. 2002. A trophic cascade regulates salt marsh primary production. Proceedings of the National Academy of Sciences (USA) 99:10500–10505.

Silvertown, J. 1982. Introduction to plant population ecology. Longman, London.

Simberloff, D. 1974. Equilibrium theory of island biogeography and ecology. Annual Review of Ecology and Systematics 5:161–182.

Simberloff, D. 1980. A succession of paradigms in ecology: essentialism to materialism and probabilism. Synthese 43:3–39.

Simberloff, D. 2004. Community ecology: is it time to move on? American Naturalist 163:787–799.

Simon, H. A. 1973. The organization of complex systems. In H. H. Pattee, editor. Hierarchy theory: the challenge of complex systems (pp. 1–27). Braziller, New York.

Skellam, J. G. 1951. Random dispersal in theoretical populations. Biometrika 38:196–218.

Slobodkin, L. B. 1985. Breakthroughs in ecology. In T. Hagerstrand, editor. The identification of progress in learning (pp. 187–195). Cambridge University Press, Cambridge.

Smith, C. 1992. How news media cover disasters: the case of Yellowstone. In P. S. Cook, editor. The future of news: television-newspapers-wire services-newsmagazines (pp. 223–240). Johns Hopkins University Press, Baltimore.

Smith, J. 2000. Nice work — but is this science? Nature 408:293.

Sober, E. 1984. The nature of selection: evolutionary theory in philosophical focus. Massachusetts Institute of Technology Press, Cambridge.

Sober, E. 1993. Philosophy of biology. Westview Press, San Francisco.

Soule, M. E., and K. A. Kohm, editors. 1989. Research priorities for conservation biology. Island Press, Washington, DC.

Sousa, W. P. 1984a. Intertidal mosaics: propagule availability, and spatially variable patterns of succession. Ecology **65**:1918–1935.

Sousa, W. P. 1984b. The role of disturbance in natural communities. Annual Review of Ecology and Systematics **15**:353–391.

Sousa, W. P. 1985. Disturbance and patch dynamics on rocky intertidal shores. In S. T. A. Pickett and P. S. White, editors. The ecology of natural disturbance and patch dynamics (pp. 101–124). Academic Press, Orlando, FL.

Srivastava, D. S. 1999. Using local-regional richness plots to test for species saturation: pitfalls and potentials. Journal of Animal Ecology **68**:1–16.

Srivastava, D. S., J. Kolasa, J. Bengtsson, A. Gonzalez, S. P. Lawler, T. E. Miller, P. Munguia, T. N. Romanuk, D. C. Schneider, and M. K. Trzcinski. 2004. Are natural microcosms useful model systems for ecology? Trends in Ecology and Evolution **19**:379–384.

Starfield, N. M., and A. L. Bleloch. 1986. Building models for conservation and wildlife management. MacMillan, New York.

Stearns, S. C. 1992. The evolution of life histories. Oxford University Press, New York.

Stegmüller, W. 1976. The structure and dynamics of theories. Springer-Verlag, New York.

Sterner, R. W. 1995. Elemental stoichiometry of species in ecosystems. In C. G. Jones and J. H. Lawton, editors. Linking species and ecosystems (pp. 240–252). Springer, New York.

Sterner, R. W., and J. J. Elser. 2002. Ecological stoichiometry: the biology of elements from molecules to the biosphere. Princeton University Press, Princeton.

Strayer, D. L., M. E. Power, W. F. Fagan, S. T. A. Pickett, and J. Belnap. 2003. A classification of ecological boundaries. BioScience **53**:723–729.

Strong, D. R. 1984. Density-vague ecology and liberal population regulation in insects. In P. W. Price, editor. A new ecology: approaches to interactive systems (pp. 313–327). Wiley, New York.

Strong, D. R., J. H. Lawton, and R. Southwood. 1984. Insects on plants: community patterns and mechanisms. Harvard University Press, Cambridge.

Suppe, F. 1977a. Afterword. In F. Suppe, editor. The structure of scientific theories (pp. 617–730). University of Illinois Press, Urbana.

Suppe, F. 1977b. Introduction. In F. Suppe, editor. The structure of scientific theories (pp. 3–5). University of Illinois Press, Urbana.

Tansley, A. G. 1935. The use and abuse of vegetational concepts and terms. Ecology **16**:284–307.

Taylor, P. F., and Y. F. Haila. 2001. Situatedness and problematic boundaries: conceptualizing life's complex ecological context. Biology & Philosophy **16**:521–532.

Taylor P. J. 2005. Unruly complexity: ecology, interpretation, engagement. University of Chicago Press, Chicago.

Thagard, P. 1992. Conceptual revolutions. Princeton University Press, Princeton.

Thompson, J. N. 1982. Interaction and coevolution. Wiley, New York.

Thompson, P. 1989. The structure of biological theories. State University of New York Press, Albany.

Thoreau, H. D. 1863. The succession of forest trees. Ticknor & Fields, Boston.

Thornley, J. H. M. 1980. Research strategy in the plant sciences. Plant and Cell Environment **33**:233–236.

Thorpe, W. H. 1974. Reductionism in biology. In F. J. Ayala, editor. Studies in the philosophy of biology: reductionism and related problems (pp. 109–138). University of California Press, Berkeley.

Tilman, D. 1982. Resource competition and community structure. Princeton University Press, Princeton.

Tilman, D. 1988. Plant strategies and the dynamics and structure of plant communities. Princeton University Press, Princeton.

Tilman, D. 1989. Ecological experimentation: strengths and conceptual problems. In G. E. Likens, editor. Long-term studies in ecology: approaches and alternatives (pp. 136–157). Springer-Verlag, New York.

Tilman, D. 1999. Diversity by default. Science **283**:495–496.

Turchin, P. 2001. Does population ecology have general laws? Oikos **94**:17–26.

Turchin, P. 2002. Does population ecology have general laws? Zhurnal Obshchei Biologii **63**:3–14.

Turchin, P. 2003. Complex population dynamics: a theoretical/empirical synthesis. Princeton University Press, Princeton.

Turner, B. L., W. C. Clark, R. W. Kates, J. F. Richards, J. T. Matthews, and W. B. Meyer, editors. 1990. The Earth as transformed by human action: global and regional changes in the biosphere over the past 300 years. Cambridge University Press, New York.

Turner, M. G. 1989. Landscape ecology: the effect of pattern on process. Annual Review of Ecology and Systematics **20**:171–197.

Turner, M. G., W. H. Romme, R. H. Gardner, R. V. O'Neill, and T. K. Kratz. 1993. A revised concept of landscape equilibrium: disturbance and stability on scaled landscapes. Landscape Ecology **8**:213–227.

Turner, J. S. 2000. The extended organism: the physiology of animal-built structures. Harvard University Press, Cambridge, MA.

Turner, W. R., T. Nakamura, and M. Dinetti. 2004. Global urbanization and the separation of humans from nature. BioScience **54**:585–590.

Ulanowicz, R. E. 1986. Growth and development: ecosystems phenomenology. Springer-Verlag, New York.

Ulanowicz, R. E. 1997. Ecology, the ascendent perspective. Columbia University Press, New York.

Ulanowicz, R. E. 2004. On the nature of ecodynamics. Ecological Complexity **1**:341–354.

Valone, T. J., and C. D. Hoffman. 2003. Population stability is higher in more diverse annual plant communities. Ecology Letters **6**:90–95.

van Fraassen, B. C. 1980. The scientific image. Oxford University Press, Oxford.

Vannote, R. L., G. W. Minshall, K. W. Cummins, J. R. Sedell, and C. E. Cushing. 1980. The river continuum concept. Canadian Journal of Fisheries and Aquatic Sciences **37**:130–137.

Vayda, A. P. 1983. Progressive contextualization: methods for research in human ecology. Human Ecology **11**:265–281.

Vermeij, G. J. 1987. Evolution and escalation: an ecological history of life. Princeton University Press, Princeton.

Vitousek, P. M. 1989. Biological invasion by *Myrica faya* in Hawaii: plant demography, nitrogen fixation, ecosystem effects. Ecological Monographs **59**:247–265.

Vitousek, P. M. 1994. Beyond global warming: ecology and global change. Ecology **75**:1861–1876.

Vitousek, P. M., and P. A. Matson. 1991. Gradient analysis of ecosystems. In J. Cole, G. M. Lovett, and S. Findlay, editors. Comparative analyses of ecosystems: patterns, mechanisms, and theories (pp. 287–298). Springer-Verlag, New York.

Wagner, F. H., and C. E. Kay. 1993. "Natural" or "healthy" ecosystems: are U.S. national parks providing them. In M. J. McDonnell and S. T. A. Pickett, editors. Humans as components of ecosystems: the ecology of subtle human effects and populated areas (pp. 257–270). Springer-Verlag, New York.

Walker, J. C. G. 1991. Biogeochemical cycles. Science **253**:686–687.

Walker, L. R., editor. 1999. Ecosystems of disturbed ground. Elsevier, New York.

Walker, L. R., and F. S. Chapin, III. 1987. Interactions among processes controlling successional change. Oikos **50**:131–135.

Waltho, N., and J. Kolasa. 1994. Organization of instabilities in multiple systems: a test of hierarchy theory. Proceedings of the National Academy of Science of the United States of America **91**:1682–1685.

Waltho, N., and J. Kolasa. 1996. Stochastic determinants of assemblage patterns in coral reef fishes: a quantification by means of two models. Environmental Biology of Fishes **47**:255–267.

Waring, G. L., and N. S. Cobb. 1992. The impact of plant stress on herbivore population dynamics. In E. Bernays, editor. Insect-plant interactions (pp. 167–226). CRC Press, Boca Raton, FL.

Watt, A. S. 1947. Pattern and process in the plant community. Journal of Ecology **35**:1–22.

Weatherhead, P. J. 1986. How unusual are unusual events? American Naturalist **128**:150–154.

Weathers, K. C., M. L. Cadenasso, and S. T. A. Pickett. 2001. Forest edges as nutrient and pollutant concentrators: potential synergisms between fragmentation, forest canopies, and the atmosphere. Conservation Biology **15**:1506–1514.

Weissman, D. 1989. Hypothesis and the spiral of reflection. State University of New York Press, Albany.

White, P. S. 1979. Pattern, process, and natural disturbance in vegetation. Botanical Review **45**:229–299.

White, P. S. 1984. The architecture of Devil's Walking Stick, *Aralia spinosa* (Araliaceae). Journal of the Arnold Arboretum Harvard University **65**:403–418.

Whittaker, R. H. 1951. A criticism of the plant association and climatic climax concepts. Northwest Science **25**:17–31.

Whittaker, R. H. 1975. Communities and ecosystems. MacMillan, New York.

Wiens, J. A. 2001. Understanding the problem of scale in experimental ecology. In R. H. Gardner, W. M. Kemp, V. S. Kennedy, and J. E. Petersen, editors. Scaling relations in experimental ecology (pp. 61–88). Columbia University Press, New York.

Williams, M. 1984. The logical status of natural selection and other evolutionary controversies. In E. Sober, editor. Conceptual issues in evolutionary biology: an anthology (pp. 83–98). MIT Press, Cambridge.

Williams, M. 1989. Americans and their forests: a historical geography. Cambridge University Press, Cambridge.

Williams, M. 1991. Agricultural impacts in temperate lands. In M. Williams, editor. Wetlands: a threatened landscape (pp. 181–206). Blackwell, Oxford.

Williams, M. B. 1970. Deducing the consequences of evolution. Journal of Theoretical Biology **29**:343–385.

Williams, R. J., and N. D. Martinez. 2000. Simple rules yield complex food webs. Nature **404**:180–183.

Wilson, E. O. 1989. Conservation: the next hundred years. In D. Western and M. C. Pearl, editors. Conservation for the twenty-first century (pp. 3–7). Oxford University Press, New York.

Wilson, E. O., and W. H. Bossert. 1971. A primer of population biology. Sinauer, Sunderland, Massachusetts.

Wilson, J. B., and W. G. Lee. 2000. C-S-R triangle theory: community-level predictions, tests, evaluation of criticisms, and relation to other theories. Oikos **91**:77–96.

Wimsatt, W. 1984. Reductionistic research strategies and their bases in the units of selection controversy. In E. Sober, editor. Conceptual issues in evolutionary biology: an anthology (pp. 142–183). MIT Press, Cambridge.

Windelband, W. 1894. History and natural science. History and Theory **19**:169–185.

Wu, J., and O. L. Loucks. 1995. From balance of nature to hierarchical patch dynamics: a paradigm shift in ecology. Quarterly Review of Biology **70**:439–466.

Wu, J., and J. L. Vankat. 1995. Island biogeography, theory and applications. In W. A. Nierenberg, editor. Encyclopedia of environmental biology (pp. 371–379). Academic Press, Orlando.

Yodzis, P. 1989. Introduction to theoretical ecology. Harper & Row, New York.

Yodzis, P. 1993. Environment and trophodiversity. In R. Ricklefs, editor. Historical and geographical determinants of community diversity (pp. 26–38). University of Chicago Press, Chicago.

Ziman, J. 1978. Reliable knowledge: an exploration of the grounds for belief in science. Cambridge University Press, Cambridge.

Ziman, J. 2000. Real science: what it is, and what it means. Cambridge University Press, New York.

Ziman, J. M. 1985. Pushing back frontiers — or redrawing maps! In T. Hagerstrand, editor. The identification of progress in learning (pp. 1–12). Cambridge University Press, New York.

INDEX

Page numbers followed by "f" denote figures; "t" denote tables; "b" denote boxes